P9-DXH-171

CALGARY PUBLIC LIBRARY

FEB    2008

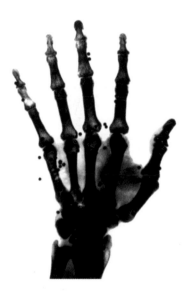

*The Story of*
# MEASUREMENT

*The Story of*
# MEASUREMENT

Andrew Robinson

With 334 illustrations, 153 in color

Thames & Hudson

For Phil Anderson,
the physicist with the highest 'creativity index', 36.9 – in admiration and gratitude

Left: *The Measurers*,
Flemish school, late
16th century. The painting
refers to instrument
making, music, weighing,
gauging, surveying,
measuring grain and
measuring cloth.

## Author's Note

This book has grown out of my research for several previous books, beginning with *The Shape of the World* in the late 1980s and culminating in *The Last Man Who Knew Everything*, a biography of Thomas Young, the 19th-century polymath and a champion measurer. Young noted of himself: 'It is probably best for mankind that the researches of some investigators should be conceived within a narrow compass, while others pass more rapidly through an extensive sphere of research.' Wisely or unwisely, I was encouraged by Young's all-embracing example to think it might be possible to tell the entire story of measurement in a brief compass. I am grateful for advice to Jim Bennett, Jonathan Bowen, Vicky Bowman, Martin Ince, Christopher Phipps, David Sprigings, Andrew Todd-Pokropek and Christopher Wood.

Metric units are used throughout, with the addition of imperial units where appropriate and the retention of the original units in some historical references.

The Story of Measurement
© 2007 Thames & Hudson Ltd, London
Text © 2007 Andrew Robinson

All Rights Reserved. No part of this publication may be reproduced or transmitted in any form or by any means, electronic or mechanical, including photocopy, recording or any other information storage and retrieval system, without prior permission in writing from the publisher.

First published in 2007 in hardcover in the United States of America by Thames & Hudson Inc., 500 Fifth Avenue, New York, New York 10110

thamesandhudsonusa.com

Library of Congress Catalog Card Number 2007921450

ISBN 978-0-500-51367-5

Printed and bound in China

p. 1: **One of the first medical radiographs made in America, in February 1896. It shows lead pellets embedded in a man's hand after a shotgun accident.**

p. 3: **The judgment of the dead from *The Egyptian Book of the Dead*, 2nd millennium BC. The heart of the deceased is weighed against the feather of *maat* (truth).**

# Contents

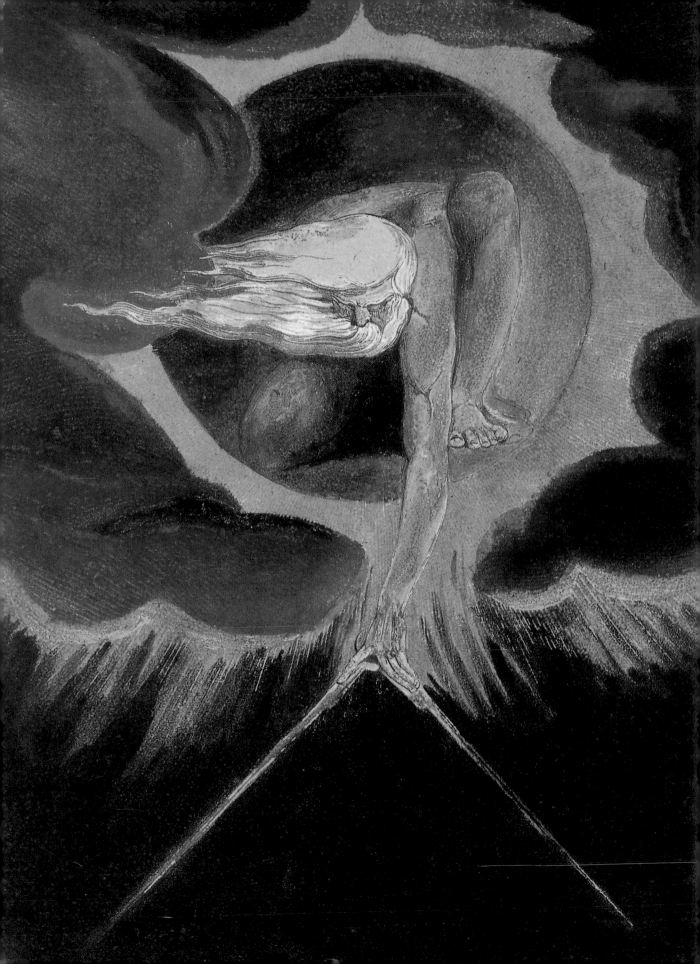

# Introduction

Throughout history, man has been ambivalent about measurement. Think of the metric system. 'Conquests may come and go, but this work will endure', a far-sighted Napoleon Bonaparte congratulated the scientists who introduced France's revolutionary new system of measurement. Yet Napoleon himself refused to use it, and the vast majority of French people followed suit. After his fall from power, Napoleon said bitterly: 'Nothing is more contrary to the organization of the mind, of the memory, and of the imagination… The new system of weights and measures will be a stumbling block and the source of difficulties for several generations.'

In general, we accept that measurement is necessary for civilization to flourish, but at the same time we feel it reduces human values to inhuman numbers – as in form-filling. The earliest known clay tablets from Mesopotamia, inscribed at the very end of the 4th millennium BC, record the hierarchy of workforces and the quantities of rations (sometimes meagre) issued to workers by the authorities. The ancient Greeks told the legend of Procrustes, a sadistic robber who fitted his victims to an iron bed by either cutting off their legs or stretching them, until he was killed by Theseus in like manner. In the Bible, there are many warnings against the sin of giving short measure, such as God's words to the prophet Micah: 'Can I connive at false scales or a bag of light weights?' Jesus (according to Luke's Gospel) extolled generosity in measures: '[G]ive, and gifts will be given to you. Good measure, pressed down, shaken together, and running over, will be poured into your lap; for whatever measure you deal out to others will be dealt to you in return.' In William Shakespeare's *Julius Caesar*, after Caesar has been stabbed to death, Antony asks: 'Are all thy conquests, glories, triumphs, spoils,/ Shrunk to this little measure?' In Charles Dickens's *Hard Times*, the successful merchant and leading citizen Thomas Gradgrind always has in his pocket 'a rule and a pair of scales', ready 'to measure any parcel of human nature, and tell you exactly what it comes to.'

A few minutes' reflection reminds us that measurement pervades our everyday lives. In no particular order, we constantly encounter: clocks, calendars, rulers, clothes sizes, floor areas, cooking recipes, sell-by dates, alcohol content, match scores, musical notation, map scales, internet protocols, word counts, memory chips, bank accounts, financial indexes, radio frequencies, calculators, speedometers, spring balances, electricity meters, cameras, thermometers, rainfall gauges, barometers, medical examinations, drug prescriptions, body mass indexes, educational tests, opinion polls, focus groups, questionnaires, consumer surveys, tax returns, censuses and many other forms of measurement – all of which serve to reduce the world to numbers and statistics. Hence, the exceptionally wide-ranging titles of the chapters in this book, from 'Atoms' and 'Mind' to 'Universe' and 'Society'.

Modern urban existence, at least in the developed world, would be unthinkable without both scientific precision and strict

*The Ancient of Days*, relief etching with watercolour by William Blake, 1794. It depicts the Creator (whom Blake called Urizen) measuring the world, and was inspired by a sentence from Proverbs in the Bible: 'I was there: when He set a compass on the face of the depths'. But since Blake, unlike the Bible, thought of the world as bad, his vision of its creation is a nightmarish one, 'in which the compass appears like a flash of lightning in a dark and stormy night' (Ernst Gombrich in *The Story of Art*). Thus Blake's famous work seems to express man's enduring ambivalence about measurement.

Ancient measuring devices for weight and length.

Left: **Cubic weights (made of banded chert or other patterned stone) from the Indus Valley were part of a system of standardized weights unique in the ancient world. The first six weights doubled in size from 1 : 2 : 4 : 8 : 16 : 32, of which the most common was the 16th ratio, weighing approximately 13.7 grams. Larger weights change to a decimal system. Amazingly, the Indus weight system is still in use in traditional markets in Pakistan and India after more than 4,000 years.**

government regulation, including taxes. 'Quantification is a way of making decisions without seeming to decide. Objectivity lends authority to officials who have very little of their own.' This shrewd comment comes from the American historian Theodore Porter in his book on measurement in public life, *Trust in Numbers*. We welcome measurement's benefits and are thrilled by its technological miracles – most recently the mobile phone, the internet, satellite navigation and the iPod. But we are also aware of the costs of measurement, and yearn to be free from its restrictions. Our very language betrays this tension. There may be 'safety in numbers', but we know that group thinking allows no freedom to be ourselves. Free-market capitalism and democracy may serve 'the greatest happiness of the greatest number', but as individuals we know there is no accounting for our tastes. Overall, we may reluctantly accept that 'our days are numbered' – the biblical 'three score years and ten' – but we still do our level best to deny the inevitability of death.

Students, for example, struggle for years to do well in school and college examinations, and their results make national newspaper headlines and provoke endless debate about educational standards. Yet innovation, creativity and leadership stubbornly refuse to be measured by such intelligence tests. An article in the leading US educational newspaper, the *Chronicle of Higher Education*, bemoaning the prevalence of multiple-choice questions in American society, concluded ruefully: 'If we are going to continue to insist on having machines grade our students, then we should expect that they are going to insist on being able to answer exam questions using the machines in their pockets' – their mobile phones with internet access.

As for their teachers, the university academics, an unstoppable growth in citation indices, journal impact factors, star ratings and league tables to measure published research, academic disciplines, university departments and universities as a whole, has produced quantity rather than quality in research, and conformity more than the hoped-for originality. 'What colour star would Socrates have got?' asked the Cambridge University philosopher Simon Blackburn in *The Times Higher Education Supplement*. 'He never wrote a thing. No measurable output at all. Rubbish.'

The same goes for our attitude to measurement in most areas of life.

Below: **The cubit ('forearm') rod from Egypt is divided into digits ('finger's breadths'), which are subdivided into fractions from one half to one sixteenth of a digit. The royal cubit, equal to 7 palms (measured without the thumb) or 28 digits, was approximately 52.3 centimetres, but other cubits, shorter and longer, were also used in Egypt; and different-sized cubits were used in ancient Babylon, Israel, Greece and Rome, varying from less than 46 cm to almost 56 cm.**

Left: **The Bronze Age sky disc of Nebra (a town in Germany), dating from perhaps 1600 BC, is thought to be the first astronomical clock. Made of bronze, the disc is inlaid with a gold sun, moon and stars. Its accuracy in keeping the solar and lunar calendars compares well with Babylonian equivalents from the 7th century BC. The disc was apparently found by illegal treasure hunters in 1999, along with other bronze objects including swords, although there is some dispute about its genuineness.**

respected statesman or leader, such as Winston Churchill or Martin Luther King, is still someone whose achievements cannot be easily measured. Psychologists who believe that intelligence, deceit and personality traits can be quantified in simple numbers like IQ or measured by instruments like lie detectors and brain scanners, attract inevitable attention. But their data and conclusions are almost always mired in controversy, especially among fellow psychologists. Physicians and surgeons who are preoccupied with using the latest, highly expensive machines and drug regimes to measure and manipulate the human body like a scientific experiment, have distorted the priorities of healthcare systems and diminished public trust in medicine. The healer, rather than the technocrat, is still the more generally admired medical practitioner.

Even among the physical scientists, the profession most susceptible to measurement (in both senses of the adjective), it is widely understood that quantification must not be

We encourage it – but suspect its results. Economists who maintain that wealth creation and human motivation can be distilled into rational economic models governed by serried mathematical equations are probably the most conspicuous of today's over-enthusiastic measurers. They multiply fruitfully, fill academic journals and win Nobel prizes, but their reputations come and go, leaving little trace in the public mind. Financiers and businessmen who think only of profit, the bottom line in the accounts, may get fabulously rich, but unlike inventors and entrepreneurs they earn limited respect. That is why so many of them convert to philanthropy in later life, funding educational, artistic and social programmes that cannot be measured by money alone. Politicians who pursue policies that will push their popularity ratings high in opinion polls may get re-elected, but as a rule they fail to change society significantly, and are forgotten or scorned once they leave office. The

Below left: **Weights and measures in *Magna Carta*, the seminal document of English constitutional practice, signed by King John in 1215 at the insistence of his barons. Translated from the medieval Latin, the relevant sentence reads: 'Let there be one measure of wine throughout our whole realm; and one measure of ale; and one measure of corn, to wit, "the London quarter"; and one width of cloth (whether dyed, or russet, or "halberget"), to wit, two ells within the selvedges; of weights also let it be as of measures.' (The English ell was about 45 inches, or 114 centimetres.)**

allowed to dominate over insight and imagination. Most things in science can be measured, with enough ingenuity – but not everything. Lord Kelvin, the 19th-century physicist whose name is attached to the absolute temperature scale, spoke for the majority of his scientific contemporaries in saying that 'when you can measure what you are speaking about and express it in numbers you know something about it, but when you cannot measure it in numbers, your knowledge is of a meagre and unsatisfactory kind.' Notwithstanding, some of the greatest physicists, including Albert Einstein and Richard Feynman, have been inspired by ideas as much as, if not more than, the numbers obtained from experimental data. 'It seems that ... knowledge cannot spring from experience alone but only from a comparison of the inventions of the intellect with the facts of observation', wrote Einstein.

## Customary Units

In the second half of the 17th century, when Isaac Newton wanted to send a letter from Cambridge to the secretary of the Royal Society in London, about fifty miles distant, he would address it *To Mr Henry Oldenburge at his house about the middle of the old Palmail in St Jameses Fields in Westminster.* No house number or postcode, indeed not even a city name. Yet the apparently vague address was considered sufficient for the purpose of correspondence before the development of London's West End.

Saharan nomads, 'for whom the exact distance from one water hole to another

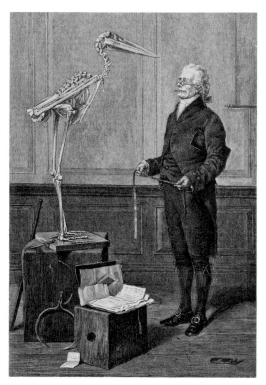

'Science is Measurement'. The phrase forms the title of this engraving from 1879–80 of a scientist armed with a tape measure who is about to examine the skeleton of a large bird. It captures a still common view of science held by scientists themselves, by social scientists and by the general public. In reality, however, while measurement is undoubtedly a necessary part of science, it is not sufficient on its own; science requires ideas and theories as well as data.

may be a matter of life or death', have long employed an extraordinary variety of terms to describe measures for long distances, notes the economic historian Witold Kula in *Measures and Men*: 'Thus, they reckon in terms of a stick's throw or a bowshot, or the carrying distance of the voice, or the distance seen with the naked eye from ground level, or from a camel's back; or walking distances from sunrise to sunset, or from early morning, mid-morning, or late morning; or a man's walking distance with no load to carry, or with a load to carry, or with a laden ass, or with an ox; or a walk across an easy or a difficult terrain.'

It is only too easy to assume that the incompatible and non-standard customary units of measurement that were the rule for most of human history must have led to endless confusion, argument and cheating. No wonder the Agora (market) in ancient Athens, as depicted in the plays of Aristophanes, resounded with noisy disputes about weights and measures. 'Uniformity of measures can only displease those lawyers who fear to see the number of lawsuits diminished, and those traders who fear a loss of profit from anything which renders commercial transactions easy and simple … A good law ought to be good for all men, as a true proposition [in geometry] is true for all men', wrote the French Enlightenment philosopher, the marquis de Condorcet, who was wholly in favour of metrication, in 1793. We moderns cannot help but ask how on earth our distant ancestors managed. How did they live with ancient units of length like the inch, the foot, the cubit and the fathom that were originally dependent on the variable size of human body parts, not to mention the diverse units of volumetric measure – gills, gallons, bushels and so on – that varied with both the type of product being measured and the country where it was being measured?

One answer is simply to point to the extraordinary enduring technological achievements of past ages. The ancient Egyptians built the Pyramids, the Greeks the Parthenon and the Romans the Colosseum and amazing aqueducts. From the past millennium, before the scientific revolution, we have structures like Angkor Wat, Chartres Cathedral and the Taj Mahal to prove that architects, engineers and craftsmen belonging to totally different civilizations could nevertheless work successfully with imprecise measures.

Another answer is to remind ourselves that earlier societies were very much more parochial than now, with far less international trade, little international communication and almost no international collaboration. Therefore monuments such as the ones just mentioned, amazing as they are, were the work of communities unified by location, language, religion and culture. They were not like today's great ventures, where different parts of a project are manufactured in various nations across the globe and then assembled on site. In earlier times, provided that all those involved on the spot agreed to use the same units of measurement, whatever these units might be, the design should interlock. No doubt they still made expensive mistakes, which architectural historians delight in spotting, but the mistakes were due not to

**Monuments before scientific measurement. The Pyramids in Egypt are among the many stupendous constructions of the pre-modern world that remind us of the practicality of customary units of measurement. Uniformity of units has become a necessity in the modern world only because of the internationalization of science and commerce.**

incompatibilities in measurement units but to avoidable misunderstandings and poor workmanship.

Yet another answer, less obvious from today's perspective, is that the concept of objective units defined by a conventionally agreed universal standard that we now take for granted, may not have suited these earlier pre-industrial societies. In Europe, for example, even as late as the end of the 18th century, it was common practice among bakers during periods of poor harvest and food shortages to keep the price of a loaf the same, but to reduce its size. This avoided a genuine difficulty arising from a lack of small coins suitable for paying for a small increase in the price; and it also preserved the 'just price' for a product recommended by the influential St Thomas Aquinas.

The adjustment in loaf size was accepted by the public, so long as it was not too obvious, in which case the plebs would riot. (The phrase, baker's dozen, to mean thirteen, perhaps derived from another solution to the problem of variable loaf sizes; the baker might throw in an extra loaf for the price of a dozen loaves.) What to us would be sharp practice and even illegal was then permitted by the authorities and public. To take another example, land measures – from the early Middle Ages until the introduction of the metric system in the 19th century – were not necessarily measures expressed in the local units of area: acres, perches, arpents or whatever. Instead, cultivable land could be measured in a wholly different way, in fact two wholly different ways: by the time taken to plough it, and by the amount of seed it required. To the peasant, and to the landowner, these numbers were actually more useful and representative than the land's geometrical area. Thus, in the pre-metric period, writes Kula, 'Measure is not a convention but a value.'

Measurement magic. Above: **Navigation using the satellites of the Global Positioning System (GPS) and** (left) **non-invasive scanning of the brain using magnetic resonance imaging (MRI) are among the most marvellous and far-reaching applications of modern measurement. They epitomize the science-fiction writer Arthur C. Clarke's third law: 'Any sufficiently advanced technology is indistinguishable from magic.'**

### The Coming of Uniformity and Precision

Despite the considerable virtues of customary units, in France by 1789 the measures had become an intolerably

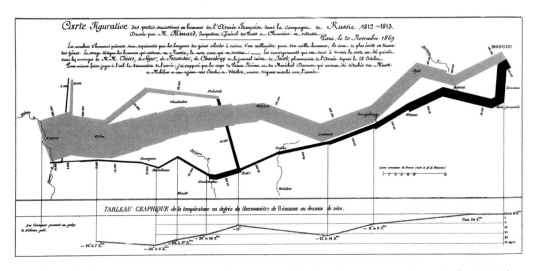

complicated drag on commerce, according to both the king and the people, as well as the scientists. The metric system was designed to replace this chaos with a single system of weights and measures, entirely decimal (except for time), based on the convention that a metre would equal one ten-millionth of the circumference of the Earth, as measured by scientist-surveyors in the 1790s. But the acute economic difficulties experienced with the new system persuaded Napoleon to rescind the original legislation in 1812 and allow the use of many of the old measures alongside the metric system, against the wishes of the scientists, who regarded this compromise as 'a retrograde decree and a bastard system'. In 1814, soon after Napoleon was deposed, the restored Bourbon monarchy, under Louis XVIII, reaffirmed this Napoleonic compromise. Not until 1837, with the post-Bourbon monarchy, was the full metric system reimposed by the French government, becoming legally binding in 1840.

During the compromise period, a British scientist sitting on a Royal Commission to consider the reform of British weights and measures made some telling observations. In the 1820s, Thomas 'Phenomenon' Young was foreign secretary of the Royal Society and superintendent of the British government's vital *Nautical Almanac*, besides being one of the greatest of Enlightenment polymaths, with strong connections among French scientists, who elected him as one of only eight foreign associates of their own scientific academy. But despite his devotion to science and to French scientists, Young did not feel able to support the legislative enactment of uniform weights and measures in Britain.

In a long historical survey on the subject written for the *Encyclopaedia Britannica* in 1823, Young noted that in France, 'it is become usual to carry in the pocket a little ruler, in the form of a triangular prism, one of the sides containing the old established lines and inches of the royal foot, a second

No safety in numbers against General Winter. According to the statistician Edward Tufte, this 1869 graphic by Charles Joseph Minard 'may well be the best statistical graphic ever drawn'. It shows the losses in Napoleon's army in its disastrous invasion of Russia in 1812, measured against the falling winter temperature on the retreat in degrees of the Réaumur thermometer below zero. (The freezing point of water is 0 °R.) The original French caption reads as follows in Tufte's translation: 'Figurative Map of the successive losses in men of the French Army in the Russian campaign 1812–1813. Drawn up by M. Minard, Inspector General of Bridges and Roads in retirement. Paris, November 20, 1869. The numbers of men present are represented by the widths of the coloured zones at a rate of one millimetre for every ten thousand men; they are further written across the zones. The brown designates the men who enter into Russia, the black those who leave it. The information which has served to draw up the map has been extracted from the works of M. M. Thiers, of Ségur, of Fezensac, of Chambray and the unpublished diary of Jacob, pharmacist of the army since October 28th.'

the millimètres, centimètres, and decimètres of the revolutionary school, and the third the new ultra-royal combination of the Jacobin measure with the royal division'. Young's conclusion, 'calmly considered', was that, rather than forcing one system on the whole country, as the French government had tried to do, the British government should 'endeavour to facilitate both the attainment of correct and uniform standards of legal existing measures of all kinds, and the ready understanding of all the provincial and local terms applied to measures, either regular or irregular, by the multiplication of glossaries and tables for the correct definition and comparison of such terms.' In other words, Young's view was that however theoretically desirable it might be to impose a common standard of scientific accuracy on all weights and measures, it was deeply undesirable to disturb the non-scientist's scale of values only for the sake of a scientific principle. Better to accommodate government to the habits of the people, than risk popular revolt against revolutionary ideology. Young would surely have agreed with Kula that units of measurement have a value that is more than merely conventional.

The same basic view prevailed in the United States of America. Thomas Jefferson, a fanatical quantifier, had favoured metrication, both during the French Revolution and later as US president, but he had given up hope of converting his countrymen. In 1821, he told John Quincy Adams, a future president who had been asked to report on the subject to the US government: 'On the subject of weights and measures, you will have, at its threshold, to encounter the question on which Solon and Lycurgus acted differently. Shall we mould our citizens to the law, or the law to our citizens?'

Neither in the United Kingdom nor in the United States did metrication become legally binding during the 19th century, as it had in France and the majority of European countries. In fact, both the UK and the US have yet to introduce metrication in full, although most parts of British life are now legally obliged to display metric units. But already in the mid-19th century the move towards metrication was evident worldwide. In 1875, in Paris, representatives of seventeen nations and empires – Argentina, the Austro-Hungarian Empire, Belgium, Brazil, Denmark, France, Germany, Italy, the Ottoman Empire, Peru, Portugal, Russia, Spain, Sweden and Norway, Switzerland, the United States of America and Venezuela (but not the United Kingdom) – signed the Convention of the Metre, 'desiring international uniformity and precision in standards of weight and measure'.

## The Système International

The 1875 convention (which was signed in 1884 by the UK) established the International Bureau of Weights and Measures at Sèvres near Paris, to be supervised by the International Committee on Weights and Measures, in turn controlled by a General Conference for Weights and Measures. Over the next many decades, these

Opposite: **The metrication of the world. Most dates refer not to the first official move towards metrication in any particular country, because such moves often faced opposition and fizzled out (as happened in Japan in the 1920s and after), but instead to the period when successful metrication of the country began in earnest. Even then, the process could take decades to complete, as it has in China since the 1920s and in the United Kingdom since the 1960s (where metrication has been exceptionally half-hearted – see pp. 30–31). Three countries (in black), the United States, Myanmar (Burma) and Liberia, have yet to enact legislation to metricate, although in the US the metric system is often employed in everyday life, especially in connection with technology and science. (The dates on the map are taken from data supplied by the US Metric Association, with some adjustments.)**

organizations progressively refined the standards for the numerous units of measurement and debated the most consistent and convenient measurement system for scientific work. In due course a consensus about units emerged. At the 11th General Conference in 1960, the Système International d'Unités, the International System of Units, generally known as the SI system, was established for science and soon became a shining example of international cooperation. The SI consists of seven basic units: the metre (for length), the kilogram (for mass), the second (for time), the ampere (for electric current), the kelvin (for thermodynamic temperature), the mole (for amount of substance) and the candela (for luminous intensity). From these are derived

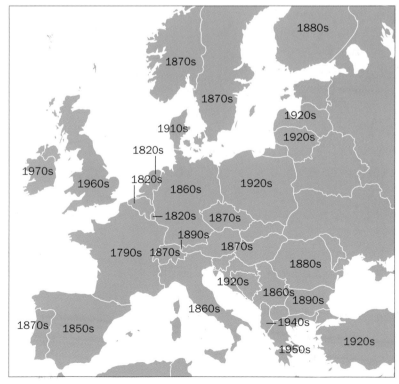

many other non-fundamental SI units, such as the hertz (for frequency), the watt (for power) and the degree Celsius (for temperature).

What brought the Convention of the Metre and then the SI into existence was, of course, the unprecedented advance in science and technology in the 19th and 20th centuries. Scientists, wherever they were located, required universally intelligible and mathematically convenient scales for measurement and calculation, scales that were capable of expressing both the minuteness of the atomic and subatomic worlds and the vastness of the astronomical world – a range extending in magnitude over more than 40 powers of ten, from quarks to galaxies. Ten to the power of minus 6 ($10^{-6}$) is a millionth, 0.000 001; ten to the power of minus 3 ($10^{-3}$) a thousandth, 0.001; ten to the power of 3 ($10^{3}$) a thousand, 1000; ten to the power of 6 ($10^{6}$) a million, 1,000,000; and so on upwards. But 10 to the power of 40 ($10^{40}$) is such an inconceivably large range of measurement that our minds boggle in trying to imagine it. The SI system, by contrast, has no difficulty in measuring such numbers, however large.

For a glimpse of the cutting edge of scientific measurement at the very smallest scale, consider a brief news report in the journal *Nature*, published in 2006. While reading it, bear in mind that the nanometre is the official SI name for $10^{-9}$ (0.000 000 001 metres), and the zeptogram the name for $10^{-21}$ (0.000 000 000 000 000 000 001 grams).

This is the news item: 'Researchers in the United States have fine-tuned a sensing technique to measure mass with zeptogram resolution. In principle, this should make it possible to weigh tens of atoms or a single molecule. ... To make the measuring device, a team of scientists at the California Institute of Technology used silicon carbide beams just 70–100 nanometres in width that vibrated at a characteristic, very high frequency. The team squirted streams of xenon and nitrogen gas at the beams and recorded the change in their vibrational frequency. Comparing the frequency shift to the quantity of gas released yielded a reliable measurement of mass to the zeptogram scale.'

The ingenuity of the measurers is amazing. However, intriguing as the idea of the zepto-world is, it does not seem to have much to do with us. Far more astonishing, surely, is one of the lesser thought experiments of Archimedes of Syracuse in an age of primitive measurements. In the 3rd century BC, this pioneering Greek scientist calculated, using his so-called Sand Reckoner, that 10 to the power of 51 ($10^{51}$), or 1,000,000,000,000,000,000,000,000, 000,000,000,000,000,000,000,000,000 grains of sand would fill the entire Universe to the outermost sphere of the fixed stars. Is man, in the end, 'the measure of all things', as first suggested by Protagoras, the most famous of the Greek Sophists, some 2,500 years ago? Or is a human being as insignificant a speck of matter in the Universe as a grain of sand on the Earth?

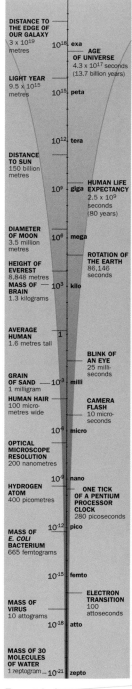

From galaxies to atoms. Science can measure over some 40 powers of ten.

# Measurement: A Chronicle

| | |
|---|---|
| Ice Ages | Lunar calendars notched on bones |
| 8000 BC | Clay 'tokens' used as counters |
| late 4th mill. BC | Measurements written on clay tablets |
| early 2nd mill. BC | Alphabet invented |
| c. 500 BC | Theory of spherical Earth |
| 3rd cent. BC | Principle of flotation discovered |
| | Earth's circumference measured |
| 129 BC | Stars catalogued |
| 46 BC | Julian calendar introduced |
| 1st century AD | Map of world with coordinates created |
| 1215 | *Magna Carta* requires uniformity in English weights and measures |
| 1543 | Heliocentric theory of solar system |
| 1582 | Gregorian calendar introduced |
| 1594 | Logarithmic tables invented |
| early 17th cent. | Optical microscope invented |
| 1609 | Telescope invented |
| 1609–20 | Laws of planetary motion |
| 1628 | Circulation of blood measured |
| 1644 | Mercury barometer invented |
| 1676 | Speed of light measured |
| 1687 | Laws of mechanics and gravitation |
| 1705 | Laws of cometary motion |
| 1720s | Mercury thermometer and Fahrenheit scale invented |
| 1735–50s | Living forms systematized |
| 1742 | Celsius scale of temperature introduced |
| 1761 | Longitude measured by marine chronometer |
| 1791 | France approves measurement standard based on meridian |
| 1792–9 | French meridian measured |
| 1801 | Metric system imposed throughout France |
| 1807 | Modulus of elasticity discovered |
| 1808 | Atomic weights and molecular formulae discovered |
| 1812 | Mohs' scale of mineral hardness introduced |
| 1816 | Stethoscope invented |
| c. 1830s | Geological timescale introduced |
| 1830s–40s | Electrical and magnetic phenomena measured; theory of electromagnetism |
| 1835 | Sphygmomanometer invented |
| 1839 | Photography invented |
| 1844 | Morse code invented |
| 1852 | World's highest mountain (Everest) measured |
| 1859 | Emission and absorption spectra measured Theory of natural selection |
| 1869 | Periodic table of elements discovered |
| 1875 | Convention of the Metre signed by 17 countries in Paris; International Bureau of Weights and Measures established World's deepest ocean trenches measured |
| 1880s | Seismograph invented Electromagnetic waves measured |
| 1884 | Greenwich meridian adopted as zero of longitude; Greenwich Mean Time (GMT) and time zones established |
| 1888 | International Phonetic Alphabet invented |
| 1890s | Fingerprinting introduced |
| 1895 | X-rays discovered |
| 1897 | Dow-Jones industrial average invented |
| 1900 | Quantum theory |
| 1900–2 | Blood types discovered |
| 1905 | Theory of special relativity |
| 1911 | Nuclear structure measured |
| 1912 | IQ measured by Stanford-Binet test Atoms measured by X-ray crystallography |
| 1913 | 'Solar system' theory of atomic structure Theory of radioactive isotopes |
| 1916 | Theory of general relativity |
| 1923 | Decibel scale introduced |
| 1927 | Heisenberg's uncertainty principle |
| 1928 | Geiger-Müller counter invented |
| 1929 | Expansion of Universe measured |
| 1931 | Transmission electron microscope invented |
| 1932 | Neutron discovered |
| 1935 | Richter earthquake magnitude introduced |
| 1940s | Electronic computers invented |
| 1950s | Caesium-beam atomic clock invented |
| 1953 | Double-helix structure of DNA discovered |
| 1957 | Sputnik, artificial satellite, launched |
| 1958 | Atmospheric carbon dioxide monitored |
| 1958–60 | Laser invented |
| 1960 | Système International (SI) introduced |
| 1960s–70s | Theory of Standard Model explains subatomic particles |
| 1965 | Cosmic background microwave radiation detected |
| 1968–72 | Apollo moonshots |
| 1969 | International Standard Book Number (ISBN) introduced |
| 1970s | Electronic calculators invented Global Positioning System (GPS) invented |
| 1975 | Fractal theory |
| 1980s | Magnetic resonance imaging (MRI) invented |
| 1990 | Hubble space telescope launched |
| 1991 | World Wide Web invented |
| 2004 | Human genome sequenced |
| 2008 | Large Hadron Collider, particle accelerator, launched |

# I THE MEANING
# OF MEASURING

Of the many things that we measure, time is the one we are most aware of in our daily lives. Everyone has access to a clock and understands the units of time: seconds, minutes, hours, days, weeks and so on. We are comfortable with 60 seconds in a minute, 60 minutes in an hour, 24 hours in a day, 7 days in a week. When the French Revolution tried to divide the day into 10 hours of 100 minutes of 100 seconds each, and establish a 10-day week, the people openly rebelled. Even the scientific Système International, a decimal system, makes the second a fundamental unit. But where did such slightly surprising numbers – 60, 24 and 7 – originate?

Ancient Mesopotamia, the Babylonians, gave us 60 minutes in an hour. The ancient Egyptians gave us 24 hours in a day, by dividing the day and the night into 12 hours each. Hellenistic astrology, combined with the Judaeo-Christian calendar, gave us the 7-day week.

Much less clear is the appeal of these particular divisions to those civilizations, and why we moderns cling to them millennia later. What is the significance of certain numbers, how did customary measures like the foot, pound and gallon evolve, and why have they meant so much to humanity?

16th-century clock tower in Bern, Switzerland. Beneath the main clock is an astronomical clock with a moving zodiac, and to its right a glockenspiel. At the very top, a male figure supposedly rings a bell with one stroke for each full hour. In the glockenspiel, a jester figure rings two bells alternately, while a roundabout of seven bears rotates, one per day of the week. This photograph was taken around 1900, when a young Swiss patent clerk in Bern, Albert Einstein, started pondering the mystery of time.

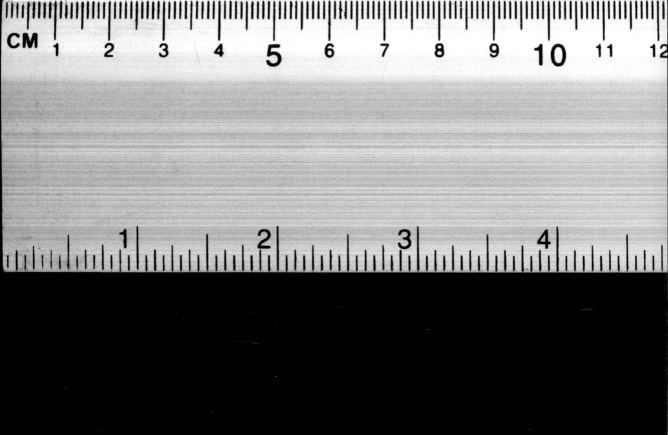

# The Measurement of the Earth by the Ancient Greeks

To the ancient Greeks, we owe two fundamental ideas about how to measure the Earth: that it is a sphere and not flat, and that the Sun revolves around it (a geocentric system) and not vice versa (a heliocentric system). The first idea was maintained by Pythagoras in the 6th century BC on theoretical grounds, and was shared by Aristotle in the 4th century on observational evidence. It had no significant dissenters. The second was accepted by all thinkers including Archimedes and Ptolemy, except for the renegade astronomer Aristarchus, who in the 3rd century proposed a rotating Earth within a heliocentric system that accounted for the seasons. The geocentric consensus would last for nearly 2,000 years until Copernicus in the 16th century.

Three later Greek thinkers, Eratosthenes, Hipparchus and Ptolemy, made further important contributions. Eratosthenes, who became director of the library at Alexandria in 235 BC, gave a remarkably accurate estimate of the Earth's circumference. Tradition has it that he used a well in far-off Syene (modern Aswan) and an obelisk in the grounds of the Alexandria library. On the day when the Sun shone directly into the well leaving no shadow on its walls – which would have been at the summer solstice since Syene is on the Tropic of Cancer – in Alexandria Eratosthenes measured the angle created between the obelisk and its shadow. Since both places lie almost on the same meridian of longitude, the angle measures the difference in latitude between Syene and Alexandria. You can see why from the diagram, in which the angle at the Earth's centre subtended by the well and the obelisk – by definition the difference in their latitudes – must, from simple geometry, be equal to the angle of the shadow. (We must assume, reasonably, that the Sun is so far away that all its rays are parallel at the Earth.) The angle was 7.2 °. Given the distance by camel from Alexandria to Syene, known to be about 5,000 *stadia*, the circumference of the Earth could be

Below: **Hipparchus observing the heavens, either in Rhodes or Alexandria.**

Bottom left: **How Eratosthenes measured the Earth's circumference. (See text for explanation.)**

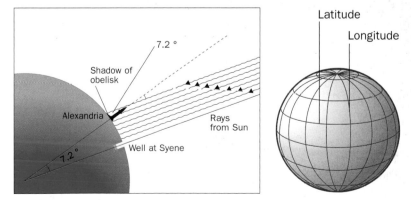

calculated by multiplying 5,000 by the ratio 360 : 7.2, giving a circumference of 250,000 *stadia*. The length of the *stadium* is disputed, but one modern conversion of this figure would give 39,690 kilometres – very close to the current value of Earth's equatorial circumference, 40,075 km.

Not a single world map survives from the classical period, but we can reconstruct the world according to Eratosthenes from surviving textual data. As is clear from the conjectural map, he did not use lines of latitude and longitude (though he put Syene on the same line as Alexandria). They were the invention of Hipparchus, a severe critic of Eratosthenes's geographical work, who lived in the 2nd century BC and compiled the first-known star catalogue, completed in 129 BC, a monumental achievement with about 850 stars measured in terms of celestial latitude and longitude and apparent brightness, using a system of six brightness magnitudes similar to that used today.

Hipparchus was a strong influence on Ptolemy, the father figure of cartography, in the 1st century AD. Ptolemy adopted Hipparchus's system of longitude, measured eastwards from a prime meridian at the Fortunate Isles (Canaries), and of latitude, measured north and south of the equator.

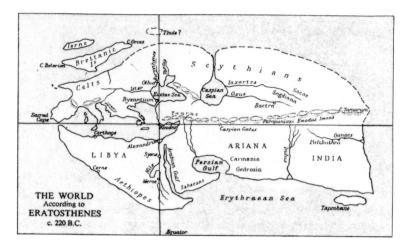

Above: **A conjectural reconstruction of Eratosthenes's map of the known world.**

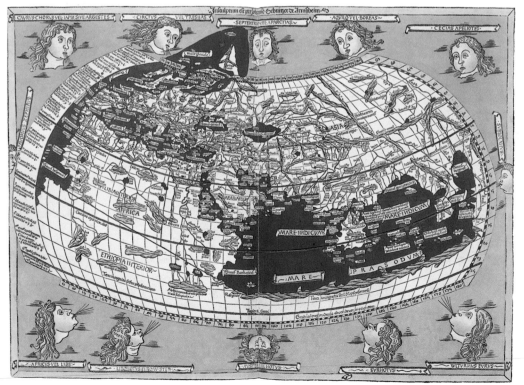

Left: **Ptolemy's map of the world, 1482. None of his original maps survives, but his gazetteer of the world, with about 8,000 places named and positioned by coordinates, was used to recreate his maps in 15th-century Italy and Germany, such as this one, published in 1482. Despite their many inaccuracies, and despite the burgeoning new geographical knowledge of the Renaissance, Ptolemy's ancient maps exerted a major influence.**

# Latitude and Longitude

Hipparchus had invented a theoretical coordinate system for specifying geographical positions on the Earth in terms of latitude and longitude. But in the long centuries of trade and exploration that followed his invention, travellers, especially seafarers, had to devise practical ways to fix their position.

The usual method at sea was 'dead reckoning', the use of a log and log-line

combined with astronomical observations. A sailor heaved overboard a piece of wood on a cord at the bow of a ship and watched it drift to the stern. By timing how long the log took to travel the length of the vessel, mariners could calculate their ship's speed (though this simple calculation took no account of sea currents). Then, by sighting the angular position of the Sun and the night-time stars with a cross-staff, back-staff or astrolabe, they could work out the ship's latitude. Hence they knew its direction of movement.

To work out one's longitude was much trickier. Ptolemy had used Pythagoras's theorem of right-angled triangles to calculate the difference in longitude between places A and B. The trouble was that the two 'known' sides of his triangle were the distance along a parallel of latitude between the latitude of A and B, and the distance between A and B as the crow flies (the hypotenuse), both of which were imprecise because of uncertainty about the size and shape of the Earth. Even assuming the Earth to be a sphere, as most thinkers now did, no one knew its exact circumference, or the length of a degree of latitude or longitude on the ground.

An alternative possibility, also known to the Greeks, involved the equation of longitude with time. As the Earth rotates on its axis, the time of sunrise gets later,

Above: **A mariner's astrolabe. Astrolabes go back to the 6th century, but their wide use among astronomers dates from the early Middle Ages in Europe and the Islamic world and from the mid-15th century among mariners. They were used to find a ship's latitude by measuring the angular height of the Sun or a star above the horizon. Astrolabes were eventually supplanted by sextants, which used telescopes for sighting** (above left).

Left: **This detail from** *The Rake's Progress* **by William Hogarth shows a lunatic drawing lines of longitude on the walls of a madhouse. Proposals to find longitude mushroomed in the 18th century.**

the further west one is from a given longitude. (In 1884, the prime meridian of longitude was fixed at Greenwich in London: New York is 5 hours behind Greenwich, San Francisco 3 hours behind New York.) The 360 degrees of longitude are encompassed during the 24 hours of the Earth's complete rotation, which means that for every 1-degree change in longitude the local time changes by 4 minutes; thus the local time on the opposite side of the globe differs by 720 minutes, or 12 hours, as we should expect.

Before the late 18th century, however, not having accurate portable clocks, navigators had no way to compare the local time – as measured by the celestial clock from detailed astronomical observations – with the time as measured in London. So longitude (like latitude) had to be determined entirely astronomically, in this instance by the method of 'lunar distances'. The increasingly

good telescopes of the 18th century allowed navigators to compare their locally measured altitude of the Moon above the horizon, the altitude of a certain star above the horizon, and the angular distance between the Moon and the Sun, with published tables of the same configurations made in London or Paris at particular times of the day and night on the same date. Since a navigator knew the local time from the heavens, he could figure his time difference with London or Paris, and hence his longitude. However, the calculation was fairly difficult, requiring correction for the effects of refraction, parallax and the dip of the horizon, and the operation was almost impossible to carry out when the deck of a ship was rolling.

For a seafaring nation like Britain, these published tables (known as ephemerides), updated regularly in the *Nautical Almanac* – started in 1767 by the astronomer royal – were of national importance well into the 19th century. But at the same time, the accuracy and portability of clocks vastly improved, beginning with the marine chronometers designed by John Harrison in the mid-18th century. After 1884, navigators could check their local time against Greenwich Mean Time by referring to a clock, and easily work out their longitude. Today, of course, both latitude and longitude are given at the press of a few buttons via the satellites of the Global Positioning System.

Below: **John Harrison (1693–1776), inventor of the marine chronometer. The son of a carpenter, Harrison devoted his long life to clock-making and built five increasingly accurate chronometers, the most famous of which is No. 4** (above). **In 1761, it was taken on trial from England to the West Indies and found to be just five seconds slow on arrival more than two months later. A copy was made for Captain James Cook on his second great voyage, and it gave Cook the confidence to begin mapping the Pacific. In 1773, after the personal intervention of King George III, Harrison was awarded the coveted prize of £20,000 offered by Parliament in 1714 for 'ascertaining the Longitude at Sea'. His No. 3 chronometer is behind him in the portrait (drawn much too large), while at his right hand lies his No. 4.**

# Figuring the Shape of the Earth

At the end of the 15th century, the ancient Greek belief that the Earth was a sphere and Ptolemy's idea of latitude and longitude were widely accepted. Map-making began to flourish, both on sheets of paper and on spherical globes. The depiction of geography was still primitive – Columbus, famously, thought that the West Indies he had discovered were islands off the east coast of Asia – but Europeans now generally understood that if they were to travel westwards or eastwards along a line of latitude, they would eventually return to their starting point.

As the geography became clearer through the great sailing journeys of the 16th century, so did the accuracy of survey techniques on land, with triangulation first reported in a book published in Antwerp in 1533. In the 17th century came telescopes with cross-hairs for sighting the stations being triangulated. These were introduced in the 1670s in France, the first country to attempt an exact survey of itself, under the direction of Jean-Dominique Cassini. The survey shifted the western coastline of existing maps about one-and-a-half degrees of longitude east in relation to the Paris meridian, and the southern coastline about half a degree of latitude to the north. Brest moved 110 miles, Marseilles 40 miles. When in 1682 King Louis XIV paid a visit to the royal observatory he had founded with a national survey uppermost in mind, and was shown the new map of his truncated kingdom, he exclaimed to the surveyors: 'Your journey has cost me a major portion of my realm!'

Across the English Channel, in Cambridge, Newton used the French data on the size of the Earth in his calculation of the force of gravity. His revolutionary new gravitational theory led to an important prediction. The Earth, said Newton, was not a perfect sphere. Centrifugal force, caused by its spinning on its axis, was balanced by gravitational force, but this did not apply evenly to the surface of the globe. The

Left: **The Paris observatory, constructed in 1667–72. Its first director was Jean-Dominique Cassini. It became Europe's leading centre of astronomy and map-making.**

Below: **One of the very earliest surviving globes, made in Nuremberg by Johannes Schöner, 1520. It is based on Martin Waldseemüller's map of 1507, and shows North and South America with a strait between them and Japan as a large island to the west of North America.**

Left: **Oblate Earth. Known as the geoid, this computer model shows the variation in gravitational attraction at different points on the Earth's surface, including a bulge at the equator due to relatively weak gravity, as first predicted by Newton's gravitational theory. On the surface of the geoid, gravitational potential is constant. The geoid shows what the Earth would look like if all the land were taken away and the water extended under land areas; it thus best represents a global mean sea level. The deviations in surface height of the geoid can be as great as 100 metres. Until the advent of the Global Positioning System, the geoid was the reference surface for determining heights around the world.**

pointed out that astronomers had observed Jupiter to be flattened at the poles. Finally, he showed how the gravitational pull of the Sun and the Moon on a bulging equator could account for the swivel in the Earth's axis of rotation that had been known to be the cause of the precession of the equinoxes since antiquity.

equator moved faster than the polar regions. The equator must therefore bulge very slightly, while the poles must be flattened, making an oblate spheroid. The Earth was like a flattened tomato. Also, gravitational attraction at the equator should be slightly less than at the poles, since Newton's theory said that gravity weakened with distance from the centre of the Earth.

To prove his case, Newton adduced some crucial evidence. First, he re-analysed the French survey of the meridian to show that the degree of latitude appeared to lengthen slightly as one moved north – an increase to be expected from a bulging equator and flattened poles. Second, Newton noted that a pendulum clock carried to the equator should beat slightly slow, since gravity was weaker there – which had indeed been observed when a French savant took such a clock to the Caribbean in 1672. Third, he

The French survey, now under the son of Cassini, was unconvinced, and maintained that its data showed the opposite: the Earth was a prolate spheroid, which was flattened at the equator and bulging at the poles. Not until the 1740s, after Newton's death, was the question resolved in Newton's favour by two gruelling expeditions led by French savants to measure a degree of latitude in Lapland, near the pole, and in Peru, at the equator. These were enough, said Voltaire, a devotee of Newton, 'to flatten both the poles and the Cassinis'.

Below left: **Jean-Dominique Cassini (1625–1712), founder of the family that mapped France. Over the century leading up to the French Revolution, four generations of Cassinis surveyed France and produced *La Carte de Cassini*, the first scientifically conducted national survey.**

Left: **Sir Isaac Newton (1642–1727), founder of modern physics, whose theory of gravitation drew on Cassini's measurements.**

# Measuring the Metre

General de Gaulle is famously supposed to have said about France: 'How can you govern a country which has 246 varieties of cheese?' Impressive though the figure is, it pales into insignificance beside the diversity of French weights and measures on the eve of the French Revolution.

Contemporary estimates suggested that under the guise of some 800 names, such as the *aune* (for length), the *arpent* (for area) and the *boisseau* (for volume), there were a staggering 250,000 measures in use in the *ancien régime*. In such a feudal system, it was quite usual for a villager to deal in three different units for the same product – one unit for the market, another for paying church tithes and yet another for paying dues to the manor. Even in the capital Paris, there reigned utter metrological confusion. The standard of the *aune* was in the charge of the guild of the Marchands Marcier in the rue Quinquempoix; the standards of measures of dry and fluid capacity were kept at the Paris town hall; the prototypes of weights were kept at the Mint and at the Grand Châtelet. In the city market, two different sets of weights from Versailles and St Denis were employed, even though a royal edict of 1778 had banned this practice. Since the time of Charlemagne in 789, eight French kings – including Louis XIV, the 'Sun King' – had attempted to lay down the law regarding standard weights and measures. All had failed, because uniformity was not in the interests of the feudal aristocracy; the lords and their stewards could generally manipulate the diversity of customary units

in their favour against their peasantry. But by 1789, even the aristocracy had realized that the chaos could not continue, such was the overwhelming demand from the people to Louis XVI in the petitions of the *cahiers de doléances*, for 'one God, one King, one law, one weight, and one measure'.

This is why the French Revolution was able to gain acceptance for such a far-reaching social change as metrication, and why France became the first country to adopt it. The people were so deeply dissatisfied with the existing muddle that they were willing to embrace a completely untried system. But they got much more than they had bargained for. In the 1790s, French scientists, backed by Republican politicians and later Napoleon, took the political opportunity to introduce an entirely decimal system (except for time) based on scientific requirements, not human habits and psychology. The conflict with the old measures was disabling, and in 1812, Napoleon – who himself still used the old units – backtracked and decreed that some of the old non-decimal fractions could continue to be used in

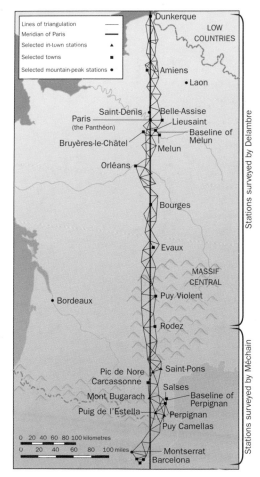

Measuring the French meridional arc. The length of the arc from Dunkerque (Dunkirk) to Barcelona was trigonometrically surveyed in 1792–9 (see pp. 108–109), and the latitude of the two end-points determined by astronomical observations, in order to calculate the length of the metre. The northern section was the work of Delambre and the southern that of Méchain.

*Lines of triangulation*
*Meridian of Paris*
*Selected in-town stations* ▲
*Selected towns* ■
*Selected mountain-peak stations* ●

Dunkerque
LOW COUNTRIES
Amiens
• Laon
Saint-Denis
Belle-Assise
Paris (the Panthéon)
Lieusaint
Baseline of Melun
Bruyères-le-Châtel
Melun
Orléans
Bourges
• Evaux
MASSIF CENTRAL
• Bordeaux
Puy Violent
Rodez
Pic de Nore
Carcassonne
Saint-Pons
Salses
Mont Bugarach
Baseline of Perpignan
Puig de l'Estella
Perpignan
Puy Camellas
Montserrat
Barcelona

0 20 40 60 80 100 kilometres
0 20 40 60 80 100 miles

Stations surveyed by Delambre

Stations surveyed by Méchain

tandem with the metric system, despite the objections of the savants. Not until 1840, half a century after the Revolution began, was the metric system finally established in France.

In the telling words of the economic historian Witold Kula, writing in *Measures and Men*: 'The reform that standardized weights and measures, which had been so ardently desired for centuries and so widely demanded by the common people on the eve of the Revolution, extolled by so many of the truest revolutionaries and conceived by the finest scientific minds of the day, had, ultimately, to be imposed upon the people.'

The very earliest proposal to relate all measurements to a decimal system based on the Earth's dimensions, rather than human or human-made dimensions, came from a French churchman, the Abbé Gabriel Mouton, writing in 1670. He suggested that the primary length unit should equal 1 minute of arc of a great circle (i.e. the circumference) of the Earth, a length not far short of 2,000 metres. The unit finally accepted by the French Academy of Sciences in 1791 was one ten-millionth of a quarter of a great circle, in other words one ten-millionth of the distance from the equator to the north pole (assuming the Earth is spherical). Since the circumference of the Earth is 40,075 kilometres (see p. 21), a quarter of this is just under 10,019 km. Dividing the latter length by 10 million gives a figure of just over 1 metre, 1.0019 m.

Prior to this, a standard based on the length of a pendulum had been considered. By the late 18th century, scientists were well

Above: **Measurers of the meridional arc. Jean-Baptiste-Joseph Delambre (1749–1822) and Pierre-François-André Méchain (1744–1804) are seen here in the uniform of the French Academy of Sciences.**

Left: **Borda's repeating circle. This trigonometrical instrument was the key to the accuracy of the seven-year meridional survey in 1792–9. Meanwhile, Jean-Charles de Borda estimated the length of a provisional metre for use in commerce, until such time as the precise metre became available from the survey of the meridional arc. His iron folding metre stick** (top left) **reads, in translation: 'Metre stick equal to one ten-millionth part of a quarter of the Earth's meridian, Borda, 1793.'**

aware that the rate of swing of a pendulum depended not at all on the weight of the bob but only on the length of the pendulum. The period of a pendulum could therefore be used to define length. In fact, the so-called seconds pendulum, which swings once every second, has a length of 0.994 m – just short of 1 m – at sea level and latitude 45 °, half way between the equator and the poles, under conditions of standard gravity. However, the pendulum measurement of the metre was rejected, partly because the period depended on gravity, which was known to vary with altitude and latitude, and partly because the units of time might themselves change. (As mentioned before, the savants were seriously considering the decimalization of time, with a 10-hour day divided into 100 minutes of 100 seconds each, though this was not ultimately accepted by the people.)

The pendulum approach might have saved a great deal of the time and trouble that was lavished on measuring as accurately as possible the length of the meridional arc stretching from Dunkirk in the north to Barcelona in the south. This challenging task, conducted at a time of turmoil in France and of war between France and Spain, took seven years and more to complete, ending in the death from malaria of one of the scientists. Its accuracy was formidable, but even then errors crept in, partly because the Earth is not a perfect sphere, partly because of error in the surveying instruments and partly because of human error in making the endless finicky observations. On the other hand, the great scientific expedition did give legitimacy to the metre and prestige to the metric system. For all its initial unpopularity, metrication was inevitable in the long run. Napoleon had been correct when he had said in 1806: 'Conquests will come and go, but this work will endure.'

## BASE

DU SYSTÈME MÉTRIQUE DÉCIMAL,

OU

### MESURE DE L'ARC DU MÉRIDIEN

COMPRIS ENTRE LES PARALLÈLES

DE DUNKERQUE ET BARCELONE,

EXÉCUTÉE EN 1792 ET ANNÉES SUIVANTES,

PAR MM. MÉCHAIN ET DELAMBRE.

Rédigée par M. Delambre, secrétaire perpétuel de l'Institut pour les sciences mathématiques, membre du bureau des longitudes, des sociétés royales de Londres, d'Upsal et de Copenhague, des académies de Berlin et de Suède, de la société Italienne et de celle de Gottingue, et membre de la Légion d'honneur.

SUITE DES MÉMOIRES DE L'INSTITUT.

TOME PREMIER.

PARIS.

BAUDOUIN, IMPRIMEUR DE L'INSTITUT NATIONAL.

JANVIER 1806.

Left: **In this French Revolutionary drawing, the citizens demonstrate (clockwise from top left) the proper use of the litre, the gram, the metre, the stere (1 cubic metre), the franc and the double-metre. The reality was in fact a great deal of confusion caused by the parallel existence of both the old and the new measures.**

Below left: **The title page of** *Base du Système Métrique Décimal*, **published in 1806. This is Delambre's own copy, in which he has written Napoleon's words to him about the book: 'Les conquêtes passent et ces opérations restent.'**

Below: **Napoleon Bonaparte (1769–1821), emperor of France 1804–14. Though keenly interested in science, Napoleon never used metric units himself. After his downfall in 1815, Napoleon attacked the savants of France for their metric obsession: 'It was not enough for them to make 40 million people happy, they wanted to sign up the whole Universe.'**

# The Metrication of the World

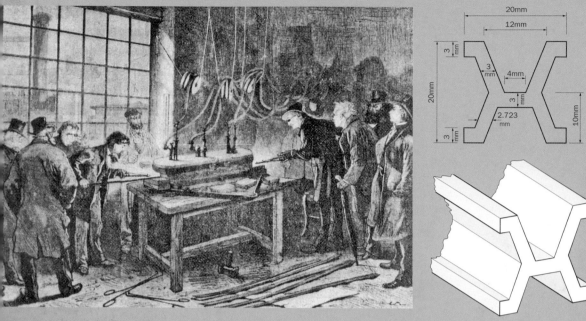

For a century and a half, from the French Revolution until the introduction of the Système International (SI) in 1960, the metre was defined by the length of a metal bar preserved in a vault at the International Bureau of Weights and Measures in Sèvres near Paris, copies of which had been distributed to national standards institutions in other countries. In 1889, a new prototype bar was made from a dense platinum-iridium alloy. It had an X-like cross-section intended to minimize sag and distortion when the bar was properly supported. On the polished facets at both ends, there were fine horizontal graticules, lines made for visual settings by micrometer, and thicker vertical lines to monitor the expansion of the metal in the temperature range 0–20 °C. The standard length was always measured at 0 °C.

The disadvantages of such a bar are obvious, and during the first half of the 20th century scientists made progressive attempts to find techniques to redefine the length of the metre in terms of the wavelength of light – an unchanging standard that could be measured in any laboratory with the right equipment. In 1960, the metre was redefined in terms of a spectrum line of krypton. Then in 1983, the current definition was adopted, based on the speed of light: the metre is now the length of the path travelled by light in a vacuum during a time of 1/299,792,458 second. This table shows how the metre's measurable accuracy has improved:

Above left: **Forging the new metre. This 1874 engraving shows scientists in the workshop of the Conservatoire Nationale des Arts et Métiers in Paris, trying to construct a new metre bar. The diagrams** above **show the end view and cross-section of the eventual platinum-iridium international metre bar.**

| Date | Basis of metre definition | Accuracy |
|------|---------------------------|----------|
| 1791 | Quarter meridian of Earth | ±0.06 mm |
| 1889 | Prototype bar | ±0.002 mm |
| 1960 | Krypton wavelength | ±0.000 007 mm |
| 1983 | Speed of light | ±0.000 000 7 mm |
| Today | Ditto, with improved laser | ±0.000 000 02 mm |

The spread of metrication around the world is shown on p. 15. After France, the next countries to metricate were its neighbours, which had come under direct French rule. Surprisingly, the metric system continued to be used in the Low Countries after the fall of Napoleon in 1815. During the compromise between old and new measures in France until 1840, Luxembourg, the Netherlands and Belgium adhered to the metric system alone.

Spain went metric in the 1850s and 60s, followed by Germany and Italy as part of their political unifications. Portugal, Norway, Sweden, Austria-Hungary and Finland soon joined in. By 1900, well over half of the European countries had gone metric. The colonial empires played their expected role. Spanish metrication meant conversion (at least officially) of Spain's remaining South American colonies, while French metrication dictated to Algeria and Tunisia, and the absence of British metrication delayed metrication in Australia, Canada and India, until the second half of the 20th century.

The first Asian nations to convert were Mongolia, in 1918, and Afghanistan and Cambodia, in the 1920s. In Japan, there was popular opposition and conversion was shelved until the 1950s; in China, metrication waited until 1959, ten years after the Communist Revolution. As in the Soviet Union, which went metric in 1924 after the Russian Revolution, political upheaval promoted metrication.

The British government officially committed itself in 1965, then dragged its feet and abolished the Metrication Board in 1979. Since 1974, the metric system has been taught in British schools, and metric packaging has been gradually introduced alongside imperial, but there is no plan to convert road signs, and press reporting uses a random mixture of imperial and metric units. Acceptance of metrication may require another decade in the UK – about the same time since 1965 as the half-century between 1791 and 1840 for France!

As for the United States, there is little political will to metricate. Even in science, the old measures are sometimes used alongside the metric system – as became embarrassingly obvious in 1999 when a Nasa probe sent to Mars was lost because one of its design teams had used the traditional units while the other had used metric. Among the US public, Gallup polls showed that between 1971 and 1991, awareness of the metric system increased from 38 to 80 per cent, but the proportion of those favouring its adoption *fell* from 50 to 26 per cent.

Metrication in the UK and the US. This low bridge sign from the UK uses four different units: yards, feet, inches and metres. In the US, awareness of the metric system is increasing, fostered by the federal government with signs such as the conversion below – but compulsory metrication is unlikely any time soon.

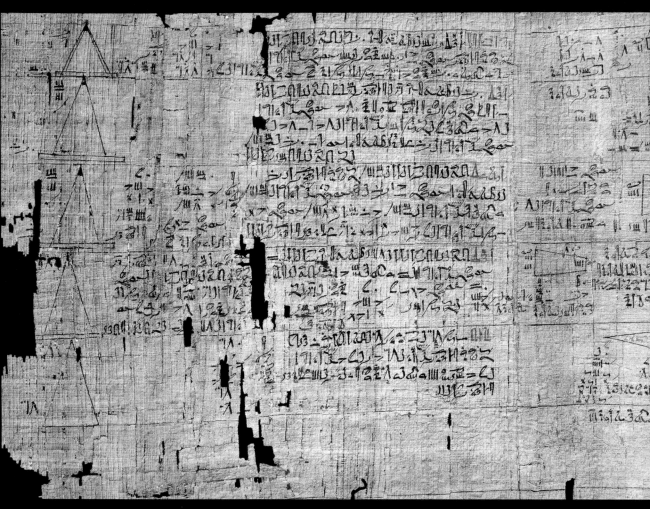

Ancient Egyptian mathematics. The Rhind mathematical papyrus dates from the 15th dynasty, *c.* 1550 BC, but claims to be a copy of a 12th-dynasty work. The series of problems shown in this part concerns the volumes of rectangles, triangles and pyramids. Another part includes an approximation for π, the about 3.1605. Babylonian mathematicians had earlier estimated π as 3.125, by calculating the perimeter of a hexagon inscribed within a circle – a geometrical method later refined by Archimedes, who estimated π as 3.1418, very close to the modern value of about 3.1416. Ancient Egyptian mathematicians were more

# Counting and Accounting

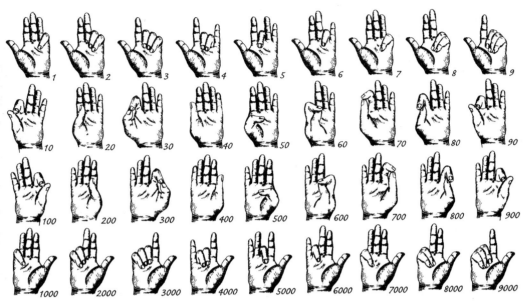

Left: **Finger symbols, from a manual published in 1520. We might imagine that our having ten fingers is the reason for our decimal system, but this idea overlooks the fact that we also have ten toes, which would have been equally available for counting in cultures where bare or sandalled feet and cross-legged sitting were the norm.**

Below: **Clay 'tokens' from the Middle East, dating from 8000–1500 BC. The crossed token seems to have been used to record numbers of sheep.**

Unlike the ability to measure, the ability to count presumes the intellectual power of abstraction. One can measure things concretely with one's finger, arm or foot – but to count them on fingers and toes requires a mental step forward: the transition from a concrete to an abstract concept. Instead of *my finger* or *your finger*, one must conceive of *the finger*. 'It must have required many ages to discover that a brace of pheasants and a couple of days were both instances of the number two', observed a perceptive Bertrand Russell.

The oldest surviving examples of counting are notched bones from the last Ice Age (see next page). The next oldest are the so-called clay 'tokens' found in excavations in the Middle East dating from 8000 BC to as late as 1500 BC, though the number of finds dated after 3000 BC falls off – a fact that has been connected with the appearance in Mesopotamia of writing on clay tablets near the end of the 4th millennium BC. The earliest tokens are undecorated and geometrically shaped – spheres, discs, cones and so on, while the later ones are often incised and shaped in more complex ways. No one knows their purpose for sure. The most probable explanation is that they were counting units in accountancy. Different shapes could have been used for different entities, such as a sheep from a flock, or a specified measure of a certain product, such as a bushel of grain. Accountancy motivated the development of all early systems and methods of counting.

# Tallies, Quipus and Abacuses

According to Herodotus, the Greek historian, the Persian king Darius left a force of Greeks (his allies) to guard a strategic bridge at his rear during an expedition against the unruly Scythians. As he was about to leave, Darius gave the Greeks a thong with sixty knots in it and told them to untie one knot each day. If he had not returned by the time all the knots were undone, he said, they should take ship for home.

The earliest known of such tallies are Ice Age bones such as these neatly notched eagle bones from Le Placard, in Charente, western France, dated to around 13,500 BC (far right). Microscopic examination suggests that the notches were made with various tools over a period of time. A plausible explanation is that the bones are lunar notations: by keeping track of the phases of the Moon, Ice Age humans created useful calendars.

Far left: **Financial tallies. In England, the Exchequer used wooden tally sticks to register the deposits and debts of its clients from the 12th century until the early 19th century. The transaction would be written on the stick, then the wood would be notched with standard marks known to all; the larger the sum, the larger the amount of wood removed. Then, the tally would be split vertically in two, and the debtor would retain the 'foil' while the creditor retained the 'stock' (hence, 'stockholder'). The metal tallies** (left), **made of gold inlay on bronze, are from China and date from the Western Zhou period, 1046–770 BC. They are apparently imitations of bamboo tallies, no examples of which survive from this time.**

CŌTADOR·MAĬORĬTE3ORERO
TAVANTĬN·SVĬO·OVĬPOC
CVRACA·CON DOR·CHAVA

con tador ytgoure          con tador

The Inca civilization of the central Andes in South America was an empire without writing (unlike the Maya and Aztec civilizations of Central America). Instead, a knotted arrangement of rope and cords called a quipu kept track of the movement of goods. Quipus were the sole bureaucratic recording device of the Incas; it was the job of *quipumayocs*, or knot keepers, in each town, to tie and interpret the knot records. The system worked well, and was retained for some time after the arrival of the Spanish conquistadores in the 16th century. It may have recorded more than simply numbers. Jeffrey Quilter remarks in *Narrative Threads*, a study of the mysteries of quipus: 'numbers may be interpreted as magnitudes or quantities, but they can also be interpreted as labels, and these labels may have narrative properties and functions.' Knot directionality, knot colours and other quipu elements are far from being fully understood.

The abacus, by contrast, has a long, widespread and familiar history as the ancestor of the modern calculating machine and computer. The derivation of the word gives a clue to its origin. It probably comes via its Greek form *abakos* from a Semitic word such as the Hebrew verb *ibeq* ('to wipe the dust') with its noun form *abaq* ('dust'). Originally, in Babylonia, the abacus was probably a board or slab with sand on it, in which calculations could be traced with the finger. Then lines and counters were added, and the lines became grooves and wires strung with the counters for rapid movement. In Europe, abacuses died out with the adoption of Arabic numerals, but they survive in the Middle East, China and Japan, where expert users can compete for speed with many modern mechanical calculating machines. The Chinese call an abacus *hsüan-pan* ('computing tray'). It uses our modern decimal place-value system.

Left: **This drawing of c. 1613 shows an Inca imperial clerk with a quipu. Its 'head' is on the left of the drawing, and its 'tail' on the right. Scholars believe that recording and reading proceeded from tail to head. There were many types of knot, each representing a value in a decimal system; the absence of a knot denoted zero. For example, a string with 2 overhand knots above a group of 4 overhand knots surmounting a 5-fold-long knot stood for the decimal number 245. The value also varied according to the knot's position in the cord.**

Below: **A Chinese abacus. In the upper section, each bead represents 5 units, and in the lower section each bead represents 1 unit. Numbers and addition operations are expressed by moving beads towards the dividing bar. The configuration shown represents the number 205,847,326,212.**

2 0 5 8 4 7 3 2 6 2 1 2

# Ancient Numerals

In one crucial respect, we still count in the way that the Sumerians counted five millennia ago. For marking time and angle, we use the sexagesimal system based on multiples of 60: there are 60 seconds in a minute, 60 minutes in an hour, 60 minutes in a degree and 360 degrees in a circle. In the Sumerian counting system, which included many subsystems, one important series of numerals was as follows:

36,000  3,600  600  60  10  1  1/2

These particular numerals were used to count most discrete objects – for example, humans and animals, dairy and textile products, fish, wooden and stone implements, and containers.

The sexagesimal system was preserved by the Babylonians in their cuneiform system, but of course the numeral signs became wedge-shaped, like all cuneiform signs. The Babylonian numerals were as follows:

$60^2 \times 10$   $60^2$   $60 \times 10$
(36,000)  (3,600)  (600)  60  10  1

By the time of the Old Babylonian period (the first half of the 2nd millennium BC), this system had developed fully. Numbers were now expressed using a place-value system, as we use today, where the value of a numeral depends on its place or position within a number. Each 5 in the decimal number 555, say, has a different value: 500,

50, 5; the place value increases by 10 times with each place, as we move from right to left. The only serious lack, 4,000 years ago, was a symbol for zero; Babylonian scribes apparently trained themselves to keep in mind an empty place within a number while calculating, where we would write a zero. (After 300 BC, they introduced a place-holder sign consisting of two slanting wedges.)

The symbols in the fully developed place-value system were as follows:

| 5 | 4 | 3 | 2 | 1 |
|---|---|---|---|---|

| 50 | 40 | 30 | 20 | 10 |
|---|---|---|---|---|

| $60^2 \times 10$ | $60^2$ | $60 \times 10$ | 60 |
|---|---|---|---|
| (36,000) | (3,600) | (600) | |

Ambiguity is immediately obvious in the use of the same symbol for 60 and 3,600; ditto for 600 and 36,000. While the numeral with the highest place value is always on the left (as in our decimal system), this still leaves three options for each of these numbers:

$60 + 10 + 5 = 75$
or $60^2 + 10 + 5 = 3,615$
or $1 + (15/60) = 1.25$

$(2 \times 60) + 40 + 5 = 165$
or $(2 \times 60^2) + (40 \times 60) + 5 = 9,605$
or $2 + (45/60) = 2.75$

In Assyrian history, there is a famous case of numerical manipulation of the place-value system. After Sennacherib sacked Babylon in 689 BC, he declared that the city must remain deserted for 70 years, by decree of the god Marduk. His son Esarhaddon, coming to the throne in 680, declared that he intended to restore Babylon, on the grounds that Marduk had relented and reversed his original number, making a curse lasting only 11 years:

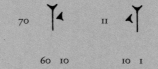

70        11

60  10        10  1

Like us (and like the Babylonians), the Maya civilization, which flourished in the period AD 250–900, used the idea of place value. But where we have a place value that increases in multiples of 10, the Maya system has a place value that increases in multiples of 20 (i.e. 1, 20, 400, 8,000 etc.). A shell symbolized zero, an advance that the Maya (and later the Indians) made over the Romans and Babylonians, and which reached Europe only much later (see p. 39). A dot stood for 1, a bar for 5. Here are some examples:

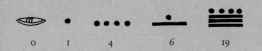

| 0 | 1 | 4 | 6 | 19 |

Instead of the place value increasing horizontally from right to left, as with our own system, among the Maya it increased vertically, moving up the page:

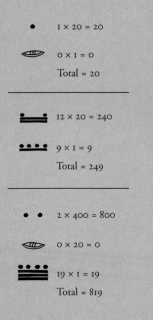

1 × 20 = 20
0 × 1 = 0
Total = 20

12 × 20 = 240
9 × 1 = 9
Total = 249

2 × 400 = 800
0 × 20 = 0
19 × 1 = 19
Total = 819

Another early civilization, that of the Minoans and Mycenaean Greeks in the mid-2nd millennium BC, used a decimal system of numerals, but without place value. In its script known as Linear B, the counting symbols are as follows:

| = 1 unit     — = 1 ten
O = 1 hundred     ⌀ = 1 thousand

Here are two examples of numbers in Linear B clay tablets, 362 and 1350:

The Greeks of the classical period, and the Romans, used a decimal system too, also without place value, but they employed letters of the alphabet as numerals. For example, in the Greek system, $\alpha$, $\beta$, $\gamma$, $\delta$ and $\epsilon$ stood for 1, 2, 3, 4 and 5; $\iota$, $\mu$ and $\pi$ for 10, 40 and 80; and $\rho$, $\tau$ and $\psi$ for 100, 300 and 700. Thus 14 would be written as $\iota\delta$, 781 as $\psi\pi\alpha$. In the Roman system, which we still use, I, V, X, L, C, D and M stood for 1, 5, 10, 50, 100, 500 and 1,000. Thus the date 1486 is MCDLXXXVI in Roman numerals.

The West's present-day numerals, 1-9, and their decimal place-value system, came to Europe in the 10th century via the court of Haroun al-Rashid, the famous caliph of Baghdad in the late 8th century, but they originated with Indian mathematicians in the 6th–7th century, so by rights should really be known as Indian, rather than Arabic, numerals. There was opposition to them – the city of Florence banned them in 1299 on the grounds that they were easier to falsify than Roman numerals – but the eventual triumph of Arabic numerals was inevitable, given the convenience and power of the system compared with all preceding numeral systems.

# Number Bases

Why did some ancient civilizations, such as the Minoan and the Egyptian, choose a decimal counting system based on 10, while others, like the Mayan and the Babylonian, selected a vigesimal system based on 20 or a sexagesimal system based on 60, respectively – or a different number base, for example duodecimal (base 12), or a mixture of number bases? The Babylonian choice of base 60 (though with some decimal elements) is especially intriguing, given its extraordinary longevity. We do not know why, despite much investigation by scholars. It seems likely that the Babylonians were influenced by the '30' days in a month and the '360' days in a year. Also important, surely, is that 60 is divisible by 30, 20, 15, 12, 10, 6, 5, 4, 3 and 2: a feature of obvious assistance in everyday transactions. (Base 12 is superior to base 10 in this respect, since 12 is divisible into halves, thirds, quarters and sixths, whereas 10 yields only halves and fifths.)

Ironically, the number base that has shaped modern civilization as much as, if not more than, base 10, was not used in the ancient world and has no helpful properties in division. Binary counting, base 2, is the basis for electronic digital computing, reflecting the fact that an electrical switch has 2 directions, off or on, as does the direction of magnetization of a spot on a magnetic disk or tape. If 'off' represents 0 and 'on' represents 1, then a series of switches – or binary digits (known as 'bits') – can easily represent any integer in a place-value notation of base 2. For an example see right.

Number bases are crucial to logarithms, invented in 1594 by John Napier (who also invented the decimal point). The common logarithm, that is the logarithm to base 10, $\log_{10}$, usually written simply as 'log', is clearest from a few examples. Recall that $10^2 = 100$, $10^3 = 1000$, $10^6 = 1,000,000$ and $10^9 = 1,000,000,000$. The logarithms of these numbers are as follows: $\log 100 = 2$, $\log 1,000 = 3$, $\log 1,000,000 = 6$ and $\log 1,000,000,000 = 9$. For intervening numbers, there are logarithms too: $\log 24 = 1.38$, $\log 759 = 2.88$ and $\log 8,525,000 = 6.94$, for example. This means that a logarithmic scale, as opposed to the linear scale we are used to, can conveniently encompass a very wide range of magnitude – a vital function in science. Logarithmic scales are used for measuring acidity and alkalinity (in pH), sound (in decibels) and earthquakes (in Richter magnitude), for instance. In a portable form, they were also the basis of the slide rule, invented in 1622, which was the scientist's chief calculator pre-electronics.

| Decimal | Binary |
|---|---|
| 0 | 0 |
| 1 | 1 |
| 2 | 10 |
| 3 | 11 |
| 4 | 100 |
| 8 | 1000 |
| 10 | 1010 |
| 32 | 100000 |
| 64 | 1000000 |
| 100 | 1100100 |

Below: **Slide rules had logarithmic scales, which were ideal for multiplication and division and the calculation of powers and roots, however large or small the numbers. The Faber-Castell 2/83N slide rule was much admired for its fine resolution and ease of use. The Apollo moonshots of 1968–72 took along slide rules as backups for their electronic calculators, but soon after this, reliable and cheap electronic calculators relegated slide rules to the museum.**

# Zero and Infinity

Speaking of the invention of the zero and the other Arabic numerals, Pierre-Simon Laplace wrote: 'It is India that gave us the ingenious method of expressing all numbers by means of ten symbols, each symbol receiving a value of position as well as an absolute value; a profound and important idea… we shall appreciate the grandeur of this achievement the more when we remember that it escaped the genius of Archimedes and Apollonius'.

While this is historically true, it begs the tricky question of whether the 'tenth' symbol, for zero, symbolizes the same kind of number as the other nine symbols? If I have five 5 coins in my pocket and pull out 3 of them, 2 coins remain in my pocket. If I take out no coins, then 5 remain. But what does it mean to say that I take 0 coins out of my pocket?

If any of the numbers 1 to 9 is added to any of those same numbers, the result is a different number, for example 2 + 2 = 4, and 7 + 8 = 15. Ditto for subtraction, division and multiplication (though you may get a negative number or a fraction). But with 0, by contrast, adding or subtracting 0 leaves a number unchanged; multiplying any number by 0 always gives the answer 0; and dividing any number by 0 presents a perplexing problem. Although we are often taught that dividing by 0 produces infinity, it would be more correct to say that dividing by 0 *does not mean anything*. You can see why if you take the simple equation $6 \times 0 = 8 \times 0$. In any equation, no matter of what kind, the familiar rule is that if you perform the same operation on each side, the equation will still balance. For example, $4 \times 12 = 6 \times 8$; divide

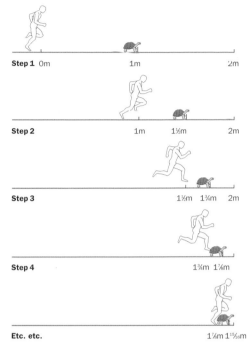

Step 1  0m          1m          2m

Step 2          1m     1½m     2m

Step 3                  1½m  1¾m  2m

Step 4                      1¾m  1⅞m

Etc. etc.                        1⅞m 1¹⁵⁄₁₆m

each side by 2, and you get $2 \times 6 = 3 \times 4$. But if we divide each side of $6 \times 0 = 8 \times 0$ by 0, we get $6 = 8$, which is nonsensical: if it were true, all numbers would be the same, since any two numbers could be substituted in the equation $a \times 0 = b \times 0$ with the same result.

The ancient Greek philosophers rejected zero – nothing, the void or vacuum – and infinity not because they were unaware of them but because these concepts appeared to be like a Trojan horse in their system of logic. Anaxagoras wrote: 'There is no smallest thing among the small and no largest among the large but always something still smaller and something still larger.' Hence the fact that Archimedes asked how many grains of sand there are in the Universe and attempted to calculate the number with his Sand Reckoner, rather than resorting to the answer: infinity.

Zeno's Second Paradox, proposed by the Greek philosopher Zeno in the 5th century BC. Better known as the paradox of Achilles and the tortoise, it posits a race in which Achilles runs at twice the speed of the tortoise but starts behind it. In the diagram, Achilles starts 1 metre behind and runs at 1 m/s and the tortoise runs at 0.5 m/s. But Achilles can never overtake the tortoise, since by the time he arrives where the tortoise was, it has already moved on. The modern solution is to accept that Achilles must take an infinite number of steps in order to catch up with the tortoise at the 2 m mark: in mathematical terms, the infinite series $1 + ½ + ¼ + ⅛ + ⅟₁₆…$ converges on the sum 2. This, and other disturbing paradoxes by Zeno, were highly influential in the Greek rejection of zero and infinity.

# Coordinates

All atlases contain coordinates. The idea goes back to Ptolemy and Hipparchus, as we know. But it was Descartes in the 17th century who introduced coordinates into mathematics. His system is still used, with its 2 or 3 axes, labelled X, Y and Z at right angles to each other, able to express position in 2 or 3 dimensions. Instead of geographical coordinates such as 51.46 °N 1.15 °W (for Oxford, England), Cartesian coordinates are expressed as, for example, (2, 4) or (2, 4, 5).

At the centre, the point where the axes cross known as the 'origin', is zero. This was a radical departure given that the ancient Greeks had rejected zero and the void or vacuum. The Catholic church, shaken by the heliocentric theory of Copernicus with its implication of an infinite cosmos, also rejected zero and the void, but there were many heretics. After all, Sylvester II, the only pope to be a published mathematician, was among those credited with the introduction of Arabic numerals (including zero) into Europe. Descartes, trained as a Jesuit, 'rejected the void but put it at the centre of his world', writes Charles Seife in *Zero*.

With Cartesian coordinates, the algebra developed by the Arabs (from Babylonian and Greek origins) could be linked with the geometry of the Greeks. Coordinates helped Einstein to discover special relativity with a simple thought experiment. You stand at the window of a railway carriage that is travelling uniformly, in other words at constant velocity, not accelerating or decelerating – and let fall a stone onto the embankment, without throwing it. If air resistance is disregarded, you, though you are moving, see the stone descend in a straight line. But a stationary pedestrian, that is someone 'at rest', who sees your action ('misdeed' writes Einstein) from the footpath, sees the stone fall in a parabolic curve. Which of the observed paths, the straight line or the parabola, is true 'in reality'? The answer is – both paths. 'Reality' here depends on which frame of reference – which system of coordinates in geometrical terms – the observer is attached to: the train's or the embankment's. One can rephrase what happens in relative terms as follows, says Einstein in *Relativity*. 'The stone traverses a straight line relative to a system of coordinates rigidly attached to the carriage, but relative to a system of coordinates rigidly attached to the ground (embankment) it describes a parabola. With the aid of this example it is clearly seen that there is no such thing as an independently existing trajectory (lit. "path-curve"), but only a trajectory relative to a particular body of reference.'

René Descartes (1596–1650), philosopher, mathematician and scientist, who invented a coordinate system that linked algebra with geometry. The diagram below left shows four points with their two-dimensional Cartesian coordinates, and four algebraic equations: two straight lines, a circle and a parabola. (The extension of the system to negative numbers was the work of Descartes's colleagues.) The equation $x = y$, plotted on Cartesian axes, is a straight line, of gradient 1, and so is $2x = y$, with a gradient of 2. Other simple equations form curves: $x^2 = y$ is a parabola; $x^2 + y^2 = 1$ is a circle, with a radius of 1.

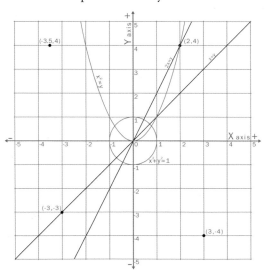

# Geometry

The inscription on the door of Plato's Academy read (in a literal translation): 'no one ungeometrical may enter'. Although the ancient Greeks undoubtedly revered the study of geometry out of pure curiosity, it had a connection with the real world. Their very word geometry means 'Earth measurement' – and geometry was certainly of immense practical importance prior to the classical period. Indeed, both geometry and arithmetic may have a common root in the surveying techniques of the pharaohs. A tightly stretched string was the Egyptian surveyor's ruler for drawing straight lines, and a pivoted string was his compass for inscribing circles.

Geometry was essential for the construction of pyramids. At the outset, it was needed to orient the axes of the sides of a pyramid to the cardinal directions. North-south was obtained by finding the direction of the noonday sun from a vertical pole set up in the sand as a gnomon (the rod in a sundial). 'Then the path traversed by the tip of its shadow would be observed. The points A and B, where this intersected a suitable circle drawn around the gnomon, would be joined. Then the line AB was bisected to establish the direction of the sun at midday', explains O. A. W. Dilke in *Mathematics and Measurement*. (Alternatively, the angle formed by the rising and setting positions of a star may have been bisected.)

The gradient of the inclined faces of pyramids – which were all square in ground plan, except for the earliest one, the step pyramid at Saqqara – depended on both the height of their apex and the length of their side. The hieroglyphic word for the gradient was *skd*, meaning ratio. It was measured in palms (7 palms = 1 cubit) by dividing half the length of a side (in palms) by the height of the apex (in cubits). There are a number of such arithmetical exercises in the Rhind mathematical papyrus (see p. 32). For example: 'A pyramid whose vertical height is 93⅓ [cubits]. Let me know its *skd*, 140 [cubits] being the length of its side.' First divide 140 by 2 to get half the length of the side, that is 70, then multiply by 7, to convert into palms, 490, then divide 490 by 93⅓, to obtain the answer 5¼ palms. Since there were 4 finger's-breadths in 1 palm, the final answer is given as 5 palms, 1 finger's-breadth. In the case of the Great Pyramid at Giza (see p. 11), which had a side of 440 cubits and an original height of 280 cubits from platform to apex, the *skd* is 5½ palms, using the same method.

The Greeks exalted geometry into realms of abstraction not conceived by the Egyptians. Euclid's five axioms and five 'common notions', first put forward in about 300 BC, dominated geometry until the 19th century and schoolbooks into the 20th century. One of the axioms – that 'A circle can be constructed when its centre, and a point on it, are given' – is still pertinent. Another, the most famous axiom, since called the parallel axiom, provoked numerous attempts to prove it until the 1820s, when yet another failure eventually gave rise to a new and important vision known as 'non-Euclidean' geometry.

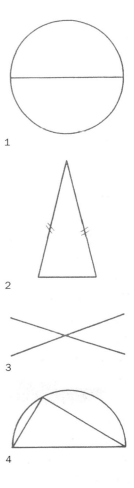

1

2

3

4

**Early Greek geometrical propositions. They are attributed to Thales of Miletus in the 6th century BC. 1 A circle is bisected by its diameter. 2 The angles at the base of an isosceles triangle are equal. 3 Two intersecting straight lines form two pairs of equal angles. 4 An angle inscribed in a semicircle is a right angle.**

# The Golden Ratio

It is often claimed that the face in Leonardo's *Mona Lisa* reveals the proportions of the golden ratio. That is, if you draw a rectangle around her face, the ratio of the rectangle's height to its width is close to 1.618 : 1 (just over 8 : 5). Leonardo never noted that the ratio was in his mind, but it is a fact that he was a close friend of Luca Pacioli, who published a three-volume treatise on the ratio, *Divina Proportione*. Pacioli particularly believed that the ratio's 'divine proportion' was to be found in the human face.

The ratio has been alleged in many other contexts, such as the Greek Parthenon, the two-column type area of Gutenberg's Bible, the spirals of the chambered nautilus shell, and even the proportions of modern credit cards. It has fascinated mathematicians ranging from Euclid (who first defined it in writing) to Kepler and Roger Penrose. 'Biologists, artists, musicians, historians, architects, psychologists, and even mystics have pondered and debated the basis of its ubiquity and appeal. In fact, it is probably fair to say that the golden ratio has inspired thinkers of all disciplines like no other number in the history of mathematics', writes the astrophysicist Mario Livio in *The Golden Ratio*. But there is no proof that any of these apparent incidences are intentional.

Mathematically, the golden ratio, now known as $\phi$ (phi) – after Phidias, the Greek sculptor said to have used it in the Parthenon – is best defined in terms of the division of a line into a larger part *a* and a smaller part *b*:

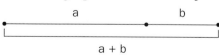

The golden ratio is the ratio when the whole line is to the larger part as the larger part is to the smaller part. To be more precise: $\phi = (a + b)/a = a/b$. It equals $(1 + \sqrt{5})/2$, an irrational number approximately equal to 1.618.

Top: **The Parthenon in Athens, 5th century BC. Two apparent instances of the golden ratio are marked on the photograph – A : B, and C : (A + B). But there is no definite evidence that the temple's Greek designers treated the golden ratio with any special respect.**

Above: **A nautilus shell. Its successive coiled chambers (which enable the shell to float) are constructed in a ratio similar to the golden ratio.**

# Fractals

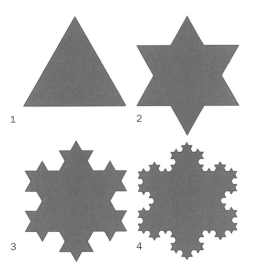

1

2

3

4

A common experience among hikers on coastal paths is that a destination is much further to walk to than the number of miles indicated on the signpost. The ancient Greeks, too, had a problem measuring the 'size' of the islands Sardinia and Sicily. Much evidence suggested that Sicily was larger than Sardinia (which we now know is marginally true in terms of area), but the sailors of antiquity contradicted the geographers. Sardinia took them longer to circumnavigate

than Sicily, because its coastline was longer. How long a coastline is turns out to be a question with no straightforward answer. A big ship will measure one figure, a small boat a larger figure, a hiker a still larger figure, and an ant presumably an even larger figure. The more we zoom in, the more fractured the coastline appears to be and the longer it becomes. The correct answer depends on how we measure the coastline.

Such roughness does not appear in the smooth geometry of Euclid, but it is an integral part of nature, and in the 20th century it led to the new geometry of fractals – a word coined from the Latin *fractus* ('broken' or 'fractured') by the mathematician Benoît Mandelbrot in 1975. In the opening words of Mandelbrot's book, *The Fractal Geometry of Nature*, 'Clouds are not spheres, mountains are not cones, coastlines are not circles, and bark is not smooth, nor does lightning travel in a straight line.'

**Fractals, natural and man-made. Each bud of cauliflower Romanesco** (below left) **looks exactly like the whole head, and can be subdivided into self-similar smaller buds over and over again. Five levels of separation can be seen with the naked eye, many more with a magnifying glass or microscope. Self-similarity at all scales of magnification is the key property of fractals, but the *degree* of self-similarity varies from the exactly similar to the qualitatively similar (as in ferns and blood vessels). Two simple mathematical fractals, the Koch snowflake** (left) **and the Sierpinski gasket** (far left), **are exactly self-similar. The snowflake is made from an equilateral triangle by iteratively replacing a line segment with four line segments creating a triangular 'bump'. (The first, second, third and fourth iterations are shown.) The length of the Koch snowflake's boundary is therefore infinite, while its area remains finite.**

Fractal analysis has in recent years been applied to the paintings of the American abstract artist Jackson Pollock, who was preoccupied with depicting what he called 'the rhythms of nature'. Known as 'Jack the Dripper', Pollock liked to work on a large canvas spread on the floor of his barn, onto which he dripped household paint using a wooden stick dipped into an old can. The results could be mysteriously appealing, as in one of his familiar works, *Autumn Rhythm*, painted in 1950. Although many have been inspired to follow Pollock's technique (a few with a view to forgery), no artist has successfully imitated the appeal of his best work. For example, the non-Pollock drip painting reproduced below.

The idea that a fractal structure might underlie Pollock's art intrigued a physicist, Richard Taylor, who is himself an abstract painter. He and colleagues subjected various Pollock paintings – and some non-Pollock drip paintings for comparison – to a

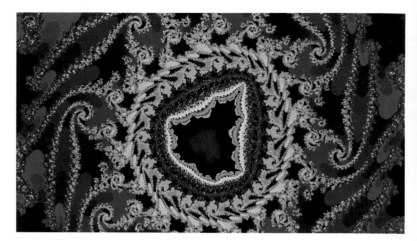

computerized analysis of pattern sizes ranging from the smallest speck of paint up to approximately a metre. They found that Pollock's patterns, but not those of the other paintings, were significantly fractal. Taylor reported in *Scientific American* in 2002: 'And they were fractal over the entire size range – the largest pattern more than 1000 times as big as the smallest.' A quarter of a century before Mandelbrot's revelation of fractals in nature, Pollock was painting them.

Above: **The fractal known as the Mandelbrot set, shown in a very small fragment, is much more complex than the Koch snowflake or the Sierpinski gasket (on the previous page), because it is *not* exactly self-similar. 'If the whole Mandelbrot set had been drawn on the same scale, the end of it would be somewhere near the star Sirius', notes Mandelbrot.**

Above: **Jackson Pollock (1912–56). The patterns in his paintings of nature are significantly fractal – unlike those of Pollock's imitators (left).**

# Mathematics: Natural or Human?

What are we to make of the fact that nature, physical reality, can be explained in terms of mathematics formulated by human beings? Do numbers and their interrelationships have a real existence 'out there', independent of the mind, which we humans 'discover', or are they pure inventions of the mind, which we impose on reality? Are there really golden ratios and fractals in nature, for example?

Ever since Galileo presented the numerical values he found in his experiments as proof that the laws of motion he had deduced were not his, but nature's, leading scientists have pondered this profound question – without coming to any conclusion. The physicist Heinrich Hertz remarked that, 'One cannot escape the feeling that these mathematical formulae have an independent existence and an intelligence of their own, that they are wiser than we are, wiser even than their discoverers, that we get more out of them than was originally put into them.' Einstein, epigrammatic as ever, said: 'As far as the propositions of mathematics refer to reality, they are not certain; and so far as they are certain, they do not refer to reality.' But later, he distinctly modified this view: 'Experience remains, of course, the sole criterion of the physical utility of a mathematical construction. But the creative principle resides in mathematics. In a certain sense, therefore, I hold it true that pure thought can grasp reality, as the ancients dreamed.'

Yet another physicist, the Nobel laureate Eugene Wigner, gave a celebrated lecture with the title, 'The unreasonable effectiveness of mathematics in the natural sciences'. There he tells the following anecdote. Two friends from high school were talking about their jobs. One of them was now a statistician studying population trends. He showed some of his published work to his former classmate. It began, as usual, with the bell-shaped curve of the Gaussian (normal) distribution and the statistician explained the meaning of the symbols for the actual population, the average population, and so on. His classmate thought this might be a joke. 'How can you know that?' he queried. 'And what is this symbol here?' 'Oh,' said the statistician, 'this is pi.' 'What is that?' 'The ratio of the circumference of the circle to its diameter.' 'Well, now you are pushing your joke too far,' said the classmate, 'surely the population has nothing to do with the circumference of the circle.'

Wigner called this reaction 'plain common sense'. He admitted: 'mathematical concepts turn up in entirely unexpected connections. Moreover, they often permit an unexpectedly close and accurate description of the phenomena in these connections… The enormous usefulness of mathematics in the natural sciences is something bordering on the mysterious and there is no rational explanation for it.'

*Circle Limit III* by M. C. Escher, 1959. Escher's classic drawings fascinate everyone, especially mathematicians, because they visualize the relationship between nature and mathematics. This example is a picture of a plane in non-Euclidean geometry. In this unfamiliar geometry of curved space, which is essential to Einstein's theory of general relativity, 'the curved lines are to be thought of as straight; all the triangles (and all the fish) are to be considered as having the same size; the bounding circle is "at infinity", and lines which meet there are parallel' (Luke Hodgkin in *A History of Mathematics: From Mesopotamia to Modernity*).

# Chapter 3 *Customary Units*

Ice Age graffiti. This stencilled hand and red dots on a boulder are probably 20,000 years old. They were made in a cave at Pech Merle, in Lot, in southern France. What does the simple but lively drawing mean? 'I was here, with my animals'? – or is the symbolism deeper? There are many similar examples in the cave, while other paintings show animals recognizable as horses, bison and mammoths, with and without dots. No one knows the meaning for sure, but it is certain that the human hand, and other body parts, such as the arm and the foot, provided the basis for the earliest systems of measuring length.

# Weight and Density

The world's earliest surviving measurement system clearly intended as such is the series of standardized stone weights of the Indus Valley civilization, dating from 2500 BC (see p. 8). Perhaps we should not be surprised that it refers to weight, rather than length, area or volume. For weight is the most important of all these measures when buying or selling food, and also precious metals like gold and silver. Hence, probably, the emphasis on weight in the familiar phrase, weights and measures; and the fact that weighing scales stand for so important a concept as justice. One of the most famous stories from antiquity – Archimedes's cry of 'Eureka!' – concerns the measurement of weight.

The pre-metric world used a huge variety of weights – from the Sumerian *shekel* and *talent* through the Greek *mina* and the Roman *libra* to the pounds and ounces of the present-day United States, not to mention such specialized weights as the scruple of the apothecaries and the carat of the jewellers.

An interesting example is the 'sack' unit used in Geneva in the 15th and 16th centuries for assessing taxes on travelling merchants. Goods were carried in pairs of sacks slung across a pack-ass. Plainly, the assessment was open to abuse by the merchant, since it depended on the number of sacks rather than their weight. He could stuff them full with heavy goods and pay the same tax as for half-empty sacks. But here a balancing mechanism came into play, tending to preserve a standard weight for the sack. 'After all,' writes Witold Kula in *Measures and Men*, it was 'primarily in [the merchant's] interest

to see that the ass should be neither too lightly loaded nor overburdened, for he might well end up on an Alpine pass with his wares strewn on the rocky ground beside his fallen beast of burden.'

Formal methods of guaranteeing standard weights – social control, supervision by the authorities and religious sanctions – were applicable to other measures, too. Social control came through the display of standards in front of the town hall or in the market, where they were easily accessible for settling arguments. Under the direction of national and local authorities, length and volume standards were cut in stone, or cast

Below: **Avoirdupois cup weight of King George III: 28 pounds, 14 pounds, 7 pounds, 4 pounds, 2 pounds, 1 pound, 8 ounces, 4 ounces, 2 ounces, 8 drams, 2 drams. One pound avoirdupois (0.454 kg) weighed 7,000 grains, the grain 'being the supposed average weight of a barley corn from the middle of the ear, equal to 64.8 mg' (R. D. Connor in *The Weights and Measures of England*). There were 16 ounces in a pound and 16 drams in an ounce. 'Avoirdupois' comes from Old French *aveir de peis* ('goods of weight').**

in heavy metal, often with ornate workmanship, to make them difficult to remove or to forge; sometimes they were riveted to a wall. Even so, they could be spirited away by the powerful during bitter battles over measures. The standards of ultimate appeal were often kept in a sacral setting such as the Jewish Temple, the Roman Capitol or the Byzantine Hagia Sophia. Today's equivalent is the International Bureau of Weights and Measures at Sèvres near Paris, which maintains the standard kilogram, and national institutions like the National Physical Laboratory in Britain.

Roman scales made of bronze from Pompeii, AD 79. The *libra* ('pound') was the standard unit of Rome, sometimes accompanied by *pondo*, making the phrase *libra pondo* ('a pound in weight'), so that *pondo* itself came to mean 'pound(s) weight'. Confusingly, 'pound' in English comes from *pondo*, but 'lb', its abbreviation, comes from *libra*. A *libra* equalled 12 *unciae* of 27–27.5 g, from which 'ounce' is derived, although there are 16 ounces in the English pound.

An Egyptian weighs gold rings against a weight in the form of a bull's head in an illustration of *c.* 1400 BC. The traditional Egyptian unit was the *deben*, weighing about 93.3 g, however this was later supplemented by the *kite* of 9–10 g, and the *deben* was fixed as equal to 10 *kite*. The *deben* measured copper, silver or gold, the *kite* only silver or gold. 'They were used to describe the equivalent value of a wide variety of non-metallic goods, thus forming a rudimentary price system in the non-monetary economy of the pharaonic period' (Ian Shaw and Paul Nicholson in the *British Museum Dictionary of Ancient Egypt*).

Checks and balances. This 15th-century stained-glass window at Tournai Cathedral in Belgium (left) shows the great beam balance weighing heavy goods. The balance was usually struck in the customer's favour on the grounds that wastage, spillage or other loss was likely. With bread, the problem was different: bakers who, by accident or design, baked underweight loaves. The penalty for the dishonest baker was to be drawn through the streets with his faulty loaf tied about his neck, and, in cases of serious offence, clapped in the stocks, as illustrated below in the *Liber Albus* (1419).

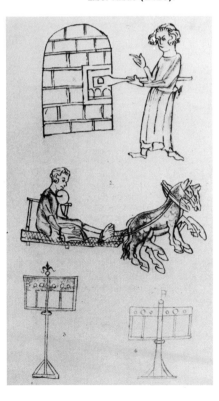

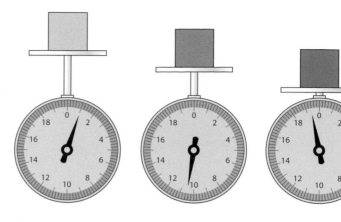

Water        Silver        Gold

Density is the ratio of weight to volume. A sack full of cement weighs far more than the same sack full of sawdust. Ice is 'lighter' than water and floats, a stone is 'heavier' than water and sinks. But it is more accurate to say a stone has a substantially higher density than water, and water a slightly higher density than ice. Gold, of course, has a much higher density than any of these. One cubic centimetre of gold weighs 19.3 grams, as compared to 10.5 g for silver and 1 g for water; the relative density of gold, compared to water, is 19.3. This is pictured in the diagram, which compares the weights of equal volumes of gold, silver and water.

Archimedes put his knowledge of weight, volume and density to practical use in Syracuse in the 3rd century BC. The ruler of the city, Hieron II, was suspicious of a crown made by his goldsmith, and asked Archimedes to find out if it was pure gold or gold alloyed with silver, without damaging it. The method is said to have occurred to Archimedes while soaking in a bathtub, when he observed the buoyancy of his body and the water it displaced. He jumped out jubilantly and ran through the streets – perhaps naked – crying 'Eureka!' meaning, more or less, 'I've got it!' The result for the unfortunate goldsmith was that he was condemned to death. Or at least so we are told by the Roman Vitruvius.

However we do not know what experiments Archimedes did. Neither do we have a statement of Archimedes's principle in his own words. A modern version is: 'The loss of weight of a body immersed in a liquid equals the weight of the liquid displaced by the body.' (An iceberg loses enough weight to float.) If two bodies have the same weight, but different volumes – different densities – the one with the larger volume (lower density) will displace more water and weigh less when immersed than the smaller one. In other words, if the crown were impure, and of lower density than a pure gold crown, it would weigh less than expected for pure gold in water. Presumably, Archimedes immersed a lump of pure gold tied to a balance and measured its loss of weight; this would have given him the relative density of gold, from which he could calculate the expected loss of weight of a pure-gold crown. Then he repeated the measurement with the actual crown, found it weighed less than calculated in water, and drew his conclusion.

Archimedes's principle, stated in the customary units of his time, would read something like 'An object fully submerged in water loses $7\tfrac{10}{19}$ *minas* of weight for every *khoes* ($\tfrac{1}{12}$ amphora) of its volume' (Alex Hebra's version in *Measure for Measure*). But although Archimedes may well have used such units, they have not survived – unlike his immortal generalization.

**Archimedes (*c.* 287–212 BC), Greek mathematician and physicist, generally considered to be the greatest in the ancient world. A parchment palimpsest containing his works, including that on floating bodies, was revealed in 2006.**

# Length and Distance

In Myanmar (Burma), a country that has yet to go metric, the traditional units of length – some of which are still in use, notably for measuring cloth – are as follows:

10 **sanchi**, a hair's breadth = 1 hnan, a sesame seed

6 **hnan** = 1 muyaw, a grain of rice

4 **muyaw** = 1 let-thit, a finger's breadth

6 **let-thit** = 1 maik, the breadth of the palm with the thumb extended

12 **let-thit** = 1 twa, the span of a hand

3 **maik or 2 twa** = 1 taung, the forearm or cubit

7 **taung** = 1 ta, a unit of land measure

1000 **ta** = 1 taing, about 2 miles

The use of the human body, and of seeds and grains, is typical of customary units of length throughout the world. Thus, in England, there was the *barleycorn*. During the early 14th century, in the reign of Edward II, the inch was defined as 'three grains of barley, dry and round, placed end to end, lengthwise.' For smaller lengths, the barleycorn was split into 4 equal parts, to create the *line*. So there were 12 lines in an inch, 12 inches (hence 36 barleycorns) in a foot, and 3 feet in a yard. However, a later rule of 1566 stated that 'foure graines of barley make a finger; foure fingers a hande; foure handes a foote.' This meant there were 64 barleycorns in a foot. Had barleycorns got smaller or feet got longer in the intervening centuries? 'The confusions and contradictions of historical unit usage defy the most ingenious present-day attempts to harmonize them or to explain them away', comments Arthur Klein in *The World of Measurements*.

Inconsistency particularly bedevilled the cubit, a unit approximately equal to the distance from a man's elbow joint to the farthest fingertip of his extended hand, or about half a metre. In the Burmese system, for example, there appear to be either 18 or 24 *let-thits* in a *taung* (cubit). In ancient Egypt, the short cubit equalled 6 palms, while the royal cubit (used in the construction of the Pyramids) equalled 7 palms. If we compare the cubits known in the early civilizations, we find the following variation in length according to A. E. Berriman's *Historical Metrology*:

| Cubit | Metric equivalent |
| --- | --- |
| Roman | 0.444 m |
| Egyptian 'short' | 0.450 m |
| Greek | 0.463 m |
| Assyrian | 0.494 m |
| Sumerian | 0.502 m |
| Egyptian 'royal' | 0.524 m |
| Talmudist | 0.555 m |
| Palestinian | 0.641 m |

Below: **The Arundel relief, Greek metrological sculpture, c. 450 BC. Its purpose is not entirely clear, but it may have been set up in a public place as a set of standard measurements. If the broken portion is completed by symmetry, then the full span of the arms, known as the fathom, measures 2.08 metres; the forearm, known as the cubit, 0.52 m; and the foot 0.297 m. However, these dimensions do not correspond to other standards of the period. For example, the Greek cubit is generally taken to be 0.463 m.**

0.297 m
0.52 m

Left: **Standardizing the foot. This woodcut from Jacob Koebel's** *Geometrie* **(1531) shows how to determine the 'right and lawful' length of the** *rute* **(about 12.36 feet) in Germany, known as the rod in England (about 16.5 feet), and also as the** *rood* **or the** *rode*, **which were similar but not identical in length. After a church service on Sunday, the surveyor should 'bid 16 men to stop, tall ones and short ones, as they happen to come out', then line them up in their Sunday best with 'their left feet one behind the other'. (The three observers in the background are probably local commissioners for weights and measures.) Koebel's figures based on the** *rute* **suggest that the average left-foot length was 9.27 inches, which seems too short, especially as the people are shown wearing shoes – so perhaps he was measuring a unit closer to the rod than the** *rute*.

The English furlong, still used in horse racing, equals one eighth of a mile. It was originally the length of a furrow for a horse ploughing a common field regarded as a square of 10 acres – hence 'furrow-long'. During the reign of Elizabeth I, a furlong was increased by statute from 625 to 660 feet (220 yards), making 8 furlongs equal to 5,280 feet, or 1,760 yards, the length of today's mile. That is about 9 per cent longer than the Roman *milliare*, which gave the mile its name. The *milliare*, too, was based on a human dimension: *mille passuum*, a 'thousand paces' of a Roman legionary on a long march. That makes each Roman 'pace' equal to about 58 inches or 1.5 metres – clearly an impossible average pace. The Roman 'pace' must therefore refer to the full cycle of left-right-left or right-left-right, giving a reasonable single pace of about 29 inches or 0.75 m.

Left: **Nilometer on Roda Island, in Cairo, completed in 861–2. The central column, octagonally shaped, was used to measure the height of the River Nile in cubits. Pre-Islamic Nilometers used stone steps as the measure.**

# Area and Volume

'Area' comes from the Latin *area*, meaning 'a vacant piece of level ground', and also a playground or threshing-floor. It shares a root with another Latin word *arere*, 'to be dry', from which comes English 'arid'. Related words are found in the modern Romance languages: *ara* in Italian, *área* in Spanish and *are* in French. The French *are* is a metric unit of area equal to 100 square metres, from which comes 'hectare', 10,000 square metres, by the addition of the prefix *hecto-*, from the Greek *hekaton*, meaning '100'.

In metric Britain, the hectare has now replaced the most important English unit of area, the ancient acre (there are just under 2.5 acres in a hectare). The acre was considered to be the amount of land ploughable by a yoke of oxen in a day. Of course this depended on such variable factors as soil, slope and drainage. Legally, an acre was a rectangular area 40 rods long by 4 rods broad. Since there were 40 rods in a furlong, an acre was also 1/8th of a mile in length by 1/80th of a mile in breadth. Hence (multiplying 1 by 8 and then 80), 1 square mile comprised 640 acres.

English measures of capacity (volume), in comparison with area, were a 'historical hodgepodge', in a telling phrase from Arthur Klein's detailed study of them. For example, under Queen Elizabeth I, 'The so-called corn gallon was restored to its former size of 268.8 cubic inches, while the wine gallon remained at 231 cubic inches. The old magnitude of the corn gallon, 282 cubic inches, became the magnitude of the new ale gallon, which was used as a measure for malt liquor.' Thus,

there were three different gallons – one for dry measure and two for liquid measure – in clear contravention of the *Magna Carta's* call, centuries earlier, for 'one weight, one measure' (see p. 9).

The most luxuriant capacity terminology applied to alcohol. In addition to the pint, quart, gallon and barrel, there were also – in ascending size – the mouthful, the jigger, the jackpot, the gill, the pottle, the peck, the bushel, the hogshead, the firkin, the pipe or butt, the kilderkin, and finally the tun. Hogsheads, like gallons, varied widely in capacity. At one time, the ale hogshead contained only 48 gallons; the beer hogshead outside London, 51 gallons; and the London beer hogshead – more generous – 54 gallons. The pipe was a double hogshead, containing 126 old wine gallons – an amount familiar to Shakespeare's jolly imbiber Falstaff from its use in the sale of sack (sherry). Sadly, such Shakespearian lines as Othello's 'Potations pottle-deep' – drink drunk to the dregs – no longer mean what they did since the disappearance of the pottle (twice a quart, half a gallon or about 2.3 litres).

The oldest surviving world map, 600 BC. It defines the Babylonian world, conceived as a flat disc surrounded by, or floating on, an ocean. The outer regions are indicated by triangles (probably eight originally), labelled in cuneiform with the distance between each region. North is at the top; Babylon is marked with an oblong, which intersects the bent parallel lines of the River Euphrates. The river flows south into a swamp, labelled by a rectangle at the mouth; a side canal may be a predecessor of the Shatt-al-Arab waterway. Dotted circles indicate cities.

Left: **This mosaic from the 'Aula dei Mensores' (Hall of the Measurers) at Ostia, an ancient port of Rome, shows slaves bringing corn from a ship and filling a** *modius*. **Usually translated as 'bushel', the** *modius* **(about 8.7 litres) was the commonest measure for dry materials. (The Roman** *amphora*, **for liquids, varied in size but held about 26.2 litres of wine or oil.)**

Below left: **Amulets shaped as the eye of the god Horus were very common in ancient Egypt; they symbolized the process of 'making whole' or healing, according to the legend of how Hathor restored the lost left eye of Horus after his fight with Seth. In this case, the eye served as a pictogram in which Egyptian scribes wrote the fractions of the** *hekat*, **the official capacity measure for corn, equal to approximately 4.8 litres; there were either 20 or 16** *hekats* **in 1** *khar* **('sack').**

Far left: **The arms of Queen Elizabeth I are visible on Winchester's standard pint of 1601. A pint equalled 32 mouthful measures, 16 handful or jigger measures, 8 jack or jackpot measures, 4 gill measures or 2 cup measures. Although most of these measures are now obsolete, the pint persists for selling milk and beer. (The British liquid pint is about 20 per cent bigger than its US equivalent.)**

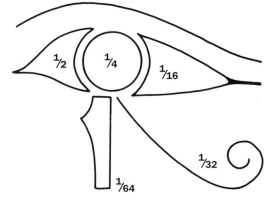

# Angle

Almost half a century ago, the astronomer Fred Hoyle made one of his radical suggestions: that the circle should be divided into 1,000 equal parts, called a *milliturn*, rather than the 360 equal parts we know as the degree. A *milliturn*, which could be further divisible into *microturns*, would equal 360/1000 = 0.36 degrees, or 21 minutes and 36 seconds of arc, since there are 60 minutes in a degree and 60 seconds in a minute. The idea got nowhere. Scientists already used an SI unit for measuring angle – the radian (see below) – and the rest of the world was too wedded to the 5000-year-old Babylonian sexagesimal system for angle (and the sexagesimal clock-face for measuring time) to consider change.

The measurement of angle did not have the same everyday importance as that of weight, length, area and volume, though it was of the highest importance to surveyors, architects and navigators, and of course astronomers. So there are no customary units of angle other than degrees, minutes and seconds. The earliest concept of angle must have come from an observer's comparing the zenith directly overhead with the far horizon on either side as seen across the flat mud-baked plains of Mesopotamia: the angular distance between zenith and horizon was clearly a right angle in either direction. It was also obvious that two right angles made up half a circle. Why the Babylonians chose 360 degrees for the whole circle we do not know, as noted before, but certainly the convenient divisibility of 360 would have helped in measuring and dividing different angles.

Over the centuries, instruments for measuring angular distance between objects developed and became more accurate. Quadrants measured a quarter of a circle (90 °), sextants a sixth of a circle (60 °), octants an eighth of a circle (45 °). At first, the surveyor or navigator sighted the distant object with his naked eye, then mirrors and telescopes were added to the instrument. In the 17th century, the telescope offered relatively little advantage over the unaided

Brass sextant of Johannes Hevelius (1611–87). The drawing is from *Machina Coelestis* (1674), and shows no telescope. When Edmond Halley visited Hevelius in Danzig in 1679 carrying telescopic sights, Halley found that Hevelius's unaided sextant, when used by its inventor, could determine stellar positions about as accurately as a telescopic sextant.

eye, but in the following century, it permitted the vastly improved accuracy found in the survey of the meridional arc during the French Revolution. However the angles measured by Borda's repeating circle (see p. 28) were not in degrees, minutes and seconds, but rather in grades. One grade was equal to 1/100th part of a right angle, hence 1/400th part of a whole circle, which was therefore divided into 400 grades not 360 degrees (1 grade equalled 0.9 degrees).

Today's Système International has the radian as its unit of angle. The concept may seem a little complicated to a non-scientist, but it makes sense, given that the circumference of a circle is $2\pi$ times its radius. There are $2\pi$ radians in any circle, and 1 radian = $360/2\pi$ = 57.296 °. More logical than dividing a circle into 360 °!

**ELIAS ALLEN.**
Apud Anglos Cantianus, iuxta **Cunnbridge** natus, Mathematici. Instrumentis ære incidendis sui temporis Artifex ingeniosissimus.
Obijt Londini, prope finem Mensis Martij. Anno a Christo nato 1653 suæque ætatis

(see p. 28)

Quadrants and their makers. During the century or so between the astronomers Tycho Brahe (1546–1601) and John Flamsteed (1646–1719), the accuracy of astronomical observations increased by a factor of 3, from about 1 minute to 20 seconds of arc. Two important instrument makers were Elias Allen (fl. 1606–54), portrayed left, and Robert Hooke (1635–1703), whose quadrant and method of screw graduation are shown below. 'Allen is holding a pair of dividers, while on the bench before him is a horizontal sundial, and an equinoctial ring, along with a circumferator. Hanging on the wall are a sector and a quadrant' (Allan Chapman in *Dividing the Circle*). In Hooke's diagram, Fig. 1 shows the complete quadrant. 'By turning this screw, the whole alidade [sighting device] and telescopic sight is advanced through a prescribed angle, as read off on the dial plate.' Fig. 11 details the screw, Fig. 13 the special catoptric sights for the instrument. Hooke's quadrant offered some improved accuracy, but nothing to compare with the 18th century, when angle measurement improved by a factor of 200.

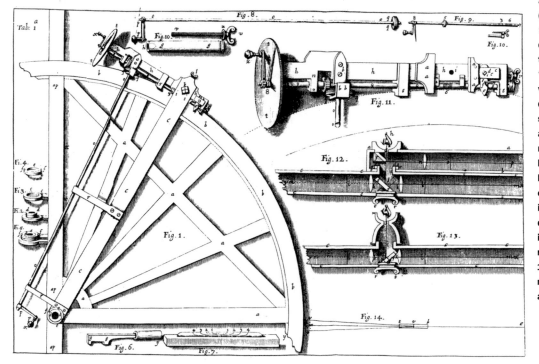

# Money and Value

'A man who knows the price of everything and the value of nothing' – Oscar Wilde's definition of a cynic – expresses an important point about the modern world. We are so habituated to the idea of money as the measure of the value of goods that we tend to forget other ways of trading. If a product's value increases, the price must increase, we assume. But instead, the price could be kept constant and the amount of goods reduced. This was once standard practice with loaves of bread, as mentioned already; moreover it was how monasteries evaded Christian restrictions on profit-making – the monks bought wine in large barrels and sold it in smaller barrels at the same price. Variable standards in weights and measures in effect acted as a currency.

But money has the key advantage that it is universally convertible, unlike goods such as grain and wine – the reason for its triumph over earlier economic arrangements. Provided everyone has confidence in money, whether it be in metal, paper or electronic form, it is a highly convenient way to compare values of goods. Furthermore, it promotes loans and other finance. A loan, in its simplest form, is an 'intertemporal value transfer', note William Goetzmann and Geert Rouwenhorst in *The Origins of Value*. 'A borrower who comes to the arrangement without any money suddenly has wealth. On the other hand, the lender takes current wealth, places it in the contractual equivalent of a time machine, and transfers it to a future date, when he might better use it.' In return, the lender typically receives interest.

A cuneiform financial contract for future delivery, 19th century BC. It records a promise to deliver wooden objects and silver. On the obverse and reverse of the inscribed tablet case, several seal impressions serve as signatures. Another cuneiform tablet (not shown) records a silver loan, *c.* 1820 BC. It names the borrower as Ilshu-bani and the lender as Sin-tajjar. The text reads: 'One and one-sixth shekels silver [i.e. 9.33 grams], to which the standard interest is to be added, Ilshu-bani, the son of Nabi-ilishu, received from [the god] Shamash and from Sin-tajjar. At harvest time he will repay the silver and the interest. Before five witnesses [their names are listed]. In month seven of the year that Apil-Sin built the temple of Inanna of Elip.'

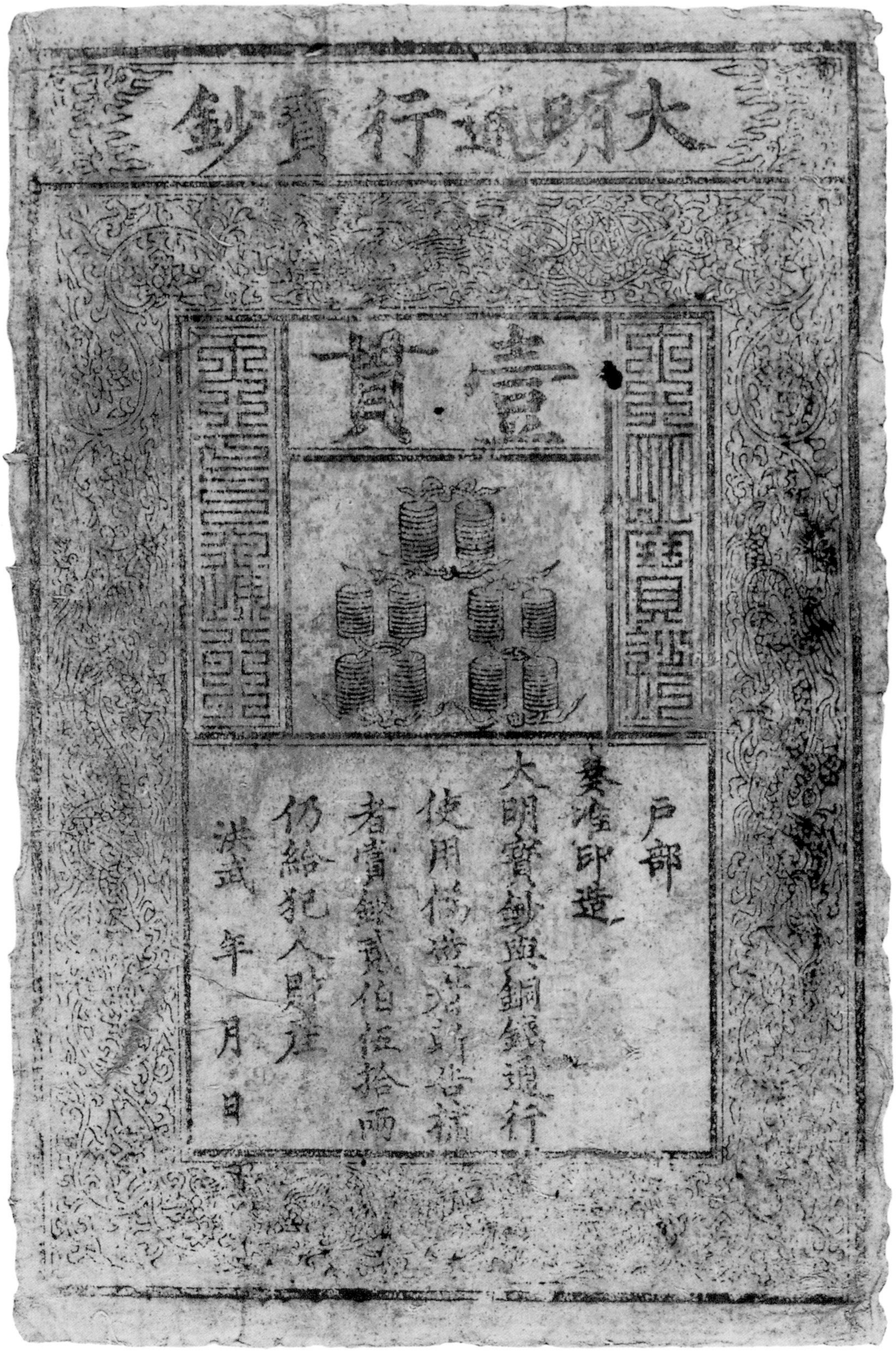

Chinese paper money. The Chinese invented paper money in the late 10th century under the Song dynasty. It suffered from periodic crises of confidence in its value, but it endured and was taken up by the Mongols under Kublai Khan, when they conquered China in the 13th century. The Ming dynasty (1368–1644) was the last imperial dynasty to use paper money. The first Ming emperor had planned to return to an all-bronze currency, but the empire's copper mines were insufficiently productive. So in 1375, the emperor introduced the paper *Da Ming tongxing baochao* ('Universally valid treasure vouchers of the Great Ming'), one of which is shown here. The text in the lower panel identifies the issuing agency and states that the bills should circulate at par with bronze coin. 'The remainder of the inscription specifies the penalty for forgery (beheading), and offers informers a reward of 250 *liang* of silver plus the property of the guilty party' (Richard Von Glahn in *The Origins of Value*). But by 1394, the value of the *baochao* had dropped to less than 20 per cent of its face value, and the emperor had to take an unprecedented step: he banned his own coinage. Nonetheless, in the 1430s paper was abandoned and silver became the dominant form of Chinese money. Paper money did not return to China until the 20th century.

N° 19 Pour 500,000 livres.

NOUS Benjamin Franklin Miniſtre Plenipotentiaire des ETATS-UNIS de l'Amerique Septentrionale, en vertu du pouvoir dont nous ſommes revetus par le CONGRES desdits Etats, promettons en ſon nom et ſolidairement pour lesdits Treize ETATS-UNIS, faire payer & rembourſer au treſor royal de ſa Majesté tres chretienne le premier Janvier mil ſept cent quatre-vingt huit au domicile de M. Grand, Banquier a Paris, la ſomme de cinque cent mille livres, argent de France, avec les interets a raiſon de cinq pour cent, l'an, valeur reçu comptant, a Paris, ce 8 Fevrier 1782.

La preſente ne devant servir que d'une ſeule & même Piece avec l'Ampliation que nous en avons delivrée aujourdhuy.

B Franklin

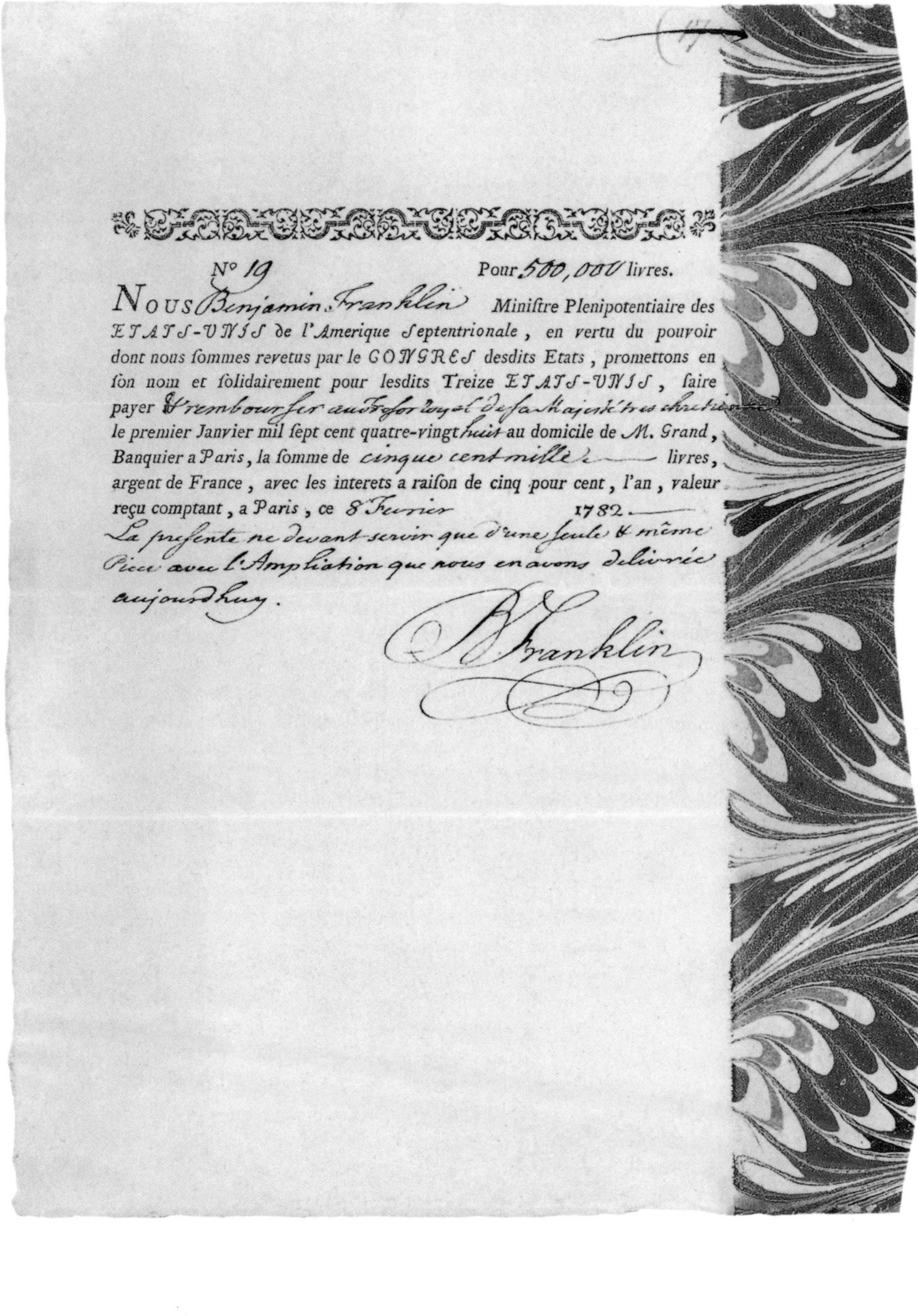

**Loans that saved a nation.** During the American Revolution, the Congress of the fledgling United States raised large sums of money from France for its war against the British. In February 1782, Benjamin Franklin, the representative of the Congress in Paris, drew up a schedule of repayment to France starting in 1788 with 12 consecutive annual payments at 5 per cent interest. As a skilled printer, Franklin printed 21 official receipts in duplicate (one for the United States, the other for France) for the individual sums, ranging from 250,000 to 3 million *livres*, which had been borrowed on 21 separate occasions. One of these receipts, for 500,000 *livres*, is shown here. These securities had a wide marbled strip down the middle that was scored to create the two matching parts of the contract – as a security against forgery. 'Ironically, Franklin procured this paper in England' (Ned Downing in *The Origins of Value*). The marbling, rarely seen in printed ephemera of this period, must have been very expensive and tricky to produce.

Cash remains vital to economic life. But the mechanical cash register – which emitted a loud *ka-ching* sound as its handle was depressed, the price of the sale popped up on tabs in a window and the closed cash drawer sprang open – is now a relic of the recent past. However, its computerized successor with laser scanner, digital readout, electronic beep and cash drawer has the same basic functions: to hold the cash securely, to issue a receipt for the customer and to turn the flow of money into a flow of data.

The first mechanical cash register was designed by a disgruntled American saloon keeper with a mechanical training. Like many retailers, James Ritty's saloon in Dayton, Ohio struggled to make a profit, even though it was busy, because he had no way of keeping accurate track of drink sales and the bartenders raided the open cash

drawer. Ritty suffered a breakdown and took a steamboat passage to Europe in 1878, hoping to recover. But on the way there, while absorbed in watching the workings of the propeller's rpm gauge below decks, he suddenly came up with the principle of the cash register. 'If the movements of a ship's propeller can be recorded,' he noted, 'there is no reason why the movements of sales in a store cannot be recorded.'

The Ritty Patent Dial Register, his first design, included a bell (the source of the *ka-ching*), but no cash drawer. The first design he marketed, as Ritty's Incorruptible Cashier, still lacked a cash drawer, but incorporated the mechanical tabs that popped up in a window. There was also an ingenious mechanism that connected the keys with a pin that punched a hole into a paper scroll according to an appropriate column – the 5-cent column, 10-cent column and so on.

An Ohio grocer, John Patterson, saw an advertisement for the contraption and bought two. He was astonished at their high cost, but even more astonished by the increase in his takings with the elimination of cash theft. In 1884, Patterson bought control of Ritty's business and renamed it the National Cash Register Company (NCR). Ritty had made further improvements, such as a customer receipt, but now came the key innovation of a cash drawer. By the turn of the century, cash registers were everywhere – and most of them were made by NCR. In 1911, the company was prosecuted by the federal government for monopolistic practices. However, the cash register helped launch the information industry and especially the success of International Business Machines (IBM).

**Both sides of a Roman *denarius*. This common coin was the origin of the abbreviation 'D' for penny in the pre-decimal British currency, used in ornate cash registers such as this one** (left) **from the turn of the 19th century. After 141 BC, the silver *denarius* was set at 1 ounce in weight and made equal to 16 *asses* – hence the 'XVI' figure on the coin – though the army could still get a *denarius* for 10 *asses*. In Britain, before decimalization in 1971, 12 pennies (D) equalled 1 shilling (/–), and 20 shillings equalled 1 pound (£).**

# Time

In Shakespeare's *Henry VI, Part 3*, the king, sitting alone on the field of battle, despairs at the horrors of war and wishes he were no better than 'a homely swain':

> *O God! methinks it were a happy life...*
> *To carve out dials quaintly, point by point,*
> *Thereby to see the minutes how they run –*
> *How many makes the hour full complete...*

When the passage of time is thus known from sundials and the seasons, the king muses, one can allocate it, and thereby spend time wisely, having planned one's life 'unto a quiet grave'.

But this division (or vision) of time is a fallacy. Time is not like weight, length, area, volume, angle, money and value. We divide it into hours and minutes as the Babylonians did, yet it moves relentlessly forwards, sometimes 'flying' and sometimes 'dragging'; and it can never be conserved like gold or geometry. 'There are no ways in which one "quantity of time" can be compared directly with another. An amount or period of time cannot be placed in storage, taken out and compared later with another', remarks Arthur Klein in *The World of Measurements*. The hands on a stopwatch, or the liquid-crystal display of a digital readout, record, not time itself, but a mechanical or electronic configuration.

What clocks and time-keepers measure is, of course, recurring cycles of motion that can be compared with changes, motions and events in the world. The measurement of time is essentially the comparison of one motion cycle with another, such as the rhythm of breathing, the revolutions of a minute hand, the repetition of night and day, or a woman's monthly period. It is hard to say whether keeping time means social or natural measurement.

Mechanical clocks date from the end of the 13th century and were driven by falling weights. The first public clock that struck the hours was erected in Milan in 1335, and the oldest surviving clock in England is the one at Salisbury Cathedral, dating from 1386. Errors per day at this time were probably as much as half an hour, and the marks depicted on the clock face were limited to the hours, at most. With the introduction of the pendulum mechanism by Christiaan Huygens in the second half of the 17th century, minute marks began to appear on the face, followed by the marking of seconds in the pendulum clocks of astronomical observatories and in marine chronometers (see p. 24) during the 18th century. In the century and a half up to 1800, accuracy improved from some 10 seconds to 1/5th of a second per day.

Since then, the measurement of time has become ever more precise. Most of us are content with the time signal on radio or television, the beeps of the 'talking clock' on the telephone or the time taken from the internet. Scientists, however, and the Système International, depend on atomic clocks (see p. 85). Like the metre, the second has now been defined ultimately in terms of nature not customary measures.

A French decimal clock, marked in Roman numerals. During the early phase of metrication in the French Revolution, a day was divided into 10 hours, each hour into 100 minutes and each minute into 100 seconds. But the change was so unpopular that it lasted less than a year, from 1794–5. A few avant-garde individuals, such as the scientist Pierre-Simon Laplace, had their watches modified, and a clock in the Palais des Tuileries kept decimal time as late as 1801. But decimal time was otherwise ignored.

The average clock and watch may have been accurate to less than a second per day in early 19th-century London, but this was of no help in keeping accurate time unless one's timepiece had been standardized against astronomical time. And that requirement meant constant interruptions for John Pond, the astronomer royal at the observatory in Greenwich, with people literally knocking on the door and asking 'Can I have a look at your clock please?', according to the current curator of horology, David Rooney. Pond eventually grew tired of being asked the time and informally deputed his assistant John Henry Belville to carry Greenwich time into the City of London and other areas of the capital on the morning of each working day.

Belville became London's first time carrier in 1836, travelling by rail and foot. The chronometer he carried had been made by the greatest British clockmaker of the day for the younger brother of King George IV, who had returned it because it was too large, 'like a warming pan'. Belville had the gold case removed and replaced with silver, fearing that he might be robbed on his rounds of London's less desirable quarters. He also changed his name to John Henry, to disguise his French origins.

The service had about 200 clients. Some were makers of timepieces, but others were banks and City firms who were increasingly conscious of the need to know the exact time of a financial transaction. There were also some private households who presumably wanted to keep Greenwich time as a status symbol. Even though a time signal sent by electric telegraph became available in 1852, Belville's service remained in demand because a telegraph line was costly to rent and not infrequently broke down. The service even survived the arrival of the time pips broadcast by radio from 1924, because the new wireless sets were expensive and required a large aerial and a licence. Then in 1936, the 'speaking clock' was made available by telephone. Carrying time by hand from Greenwich was finally discontinued after a century in 1939, still with fifty subscribers. Throughout this period, the service had been kept in the family – first Belville, then his widow and finally, from 1892, their daughter Ruth, who is shown here calling at the Greenwich observatory about a century ago, to note GMT on its 24-hour clock.

**Time carrier. Ruth Belville calls at the Greenwich observatory before carrying Greenwich Mean Time (GMT) by clock all over London.**

An elephant water clock, based on 13th-century drawings by Ibn al-Razzaz al-Jazari, and a working model from Dubai (above). It depends on a water-filled bucket concealed inside the elephant, in which floats a deep bowl with a small hole in its centre. The bowl takes exactly 30 minutes to fill through the hole. As it sinks, it pulls a string attached to a 'see-saw' in the tower, which releases a ball that drops into the mouth of a snake. This causes the snake to tip forward while a system of strings makes the figure in the tower raise either his right or left hand and the mahout strike his drum, to mark the full or the half hour. The snake now tips back again and as it does so, the sunken bowl is hauled up – and the cycle begins again.

# Chapter 4 *Instruments and Techniques*

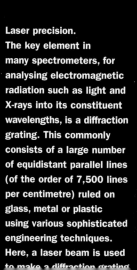

**Laser precision.**
**The key element in
many spectrometers, for
analysing electromagnetic
radiation such as light and
X-rays into its constituent
wavelengths, is a diffraction
grating. This commonly
consists of a large number
of equidistant parallel lines
(of the order of 7,500 lines
per centimetre) ruled on
glass, metal or plastic
using various sophisticated
engineering techniques.
Here, a laser beam is used
to make a diffraction grating.**

# Accuracy and Precision, Error and Uncertainty

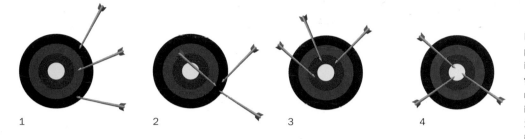

1      2      3      4

The great meridional survey of France in the 1790s was precise but not accurate. This statement may surprise and perhaps confuse, but that is because speech and writing use the words accuracy and precision almost interchangeably; indeed the *Oxford English Dictionary* defines 'accuracy' as 'exactness or precision' and 'precision' as 'accuracy'! Yet there is a significant difference (see above), and for scientists the distinction is important.

In the French survey (see pp. 27–9), the distances measured by Méchain in the southern portion of the arc were precise because they were *repeatable* and therefore internally consistent, but they were not accurate because they did not correspond well with the distances measured at the time by Delambre in the northern portion and subsequently confirmed by others. The main reason for the inaccuracy, which Méchain did not fully appreciate, was the irregularity of the Earth's crust and gravity near the Pyrenees, which distorted the arc and affected his surveying instrument. In addition, an error may have crept in through wear and tear on Borda's repeating circle caused by its constant manipulation. So vexed and ashamed was the fastidious Méchain, that in the end he fudged his data, as discovered by a disturbed but forgiving Delambre after his colleague's death from malaria.

'Error' and 'uncertainty', too, must be distinguished by careful measurers. The definition of the UK's National Physical Laboratory is: 'Error is the difference between the *measured value* and the *true value* of the object being measured. Uncertainty is the quantification of the doubt about the measurement result.' If we say that a metal rod is '300 centimetres long, plus or minus 0.1 per cent, at a 95 per cent confidence level', we mean that we are 95 per cent sure that its length lies between 299.7 and 300.3 cm. Errors, such as a bias in a spring balance that always underweighs by 50 grams, can in principle be eliminated from a measurement result, but uncertainty can never be entirely removed, only estimated on the basis of all possible sources of inaccuracy and then stated as a degree of confidence.

**Left: Accuracy versus precision. The bull's-eye in the target represents the true value of a measurement. 1 is inaccurate, and imprecise; 2 is precise but inaccurate; 3 is accurate but imprecise; and 4 is accurate and precise.**

**Below: Traceability. This pyramid shows how the uncertainty in a specification increases as we pass from a national standards laboratory to a manufactured product in the shops. By calibrating an instrument against a more accurate instrument, the first instrument's accuracy of measurement can be linked or 'traced' back up the chain to the national standard. 'This demonstrable linkage to national standards with known accuracy is called *traceability*', in the words of the UK's National Physical Laboratory.**

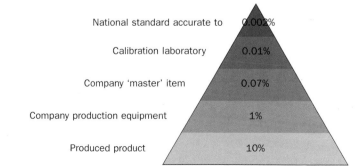

| National standard accurate to | 0.002% |
| Calibration laboratory | 0.01% |
| Company 'master' item | 0.07% |
| Company production equipment | 1% |
| Produced product | 10% |

# How Long is a Piece of String?

How long does it take to travel across New York, or to write a book? Somebody might reply: 'How long is a piece of string?' In other words, the answer depends radically on definitions and circumstances. One could provide a figure, but its degree of uncertainty would make it meaningless. A piece of string can be any length one chooses, depending on where one cuts the ball of string.

Suppose we measure the length of a particular piece. Even then, our result will have some uncertainty, depending on factors ranging from the ruler we use, through how careful we are in the measuring, to environmental conditions like humidity and temperature. Indeed, all measurements suffer from similar uncertainties. Measuring instruments may have a bias, may change with age and wear, and may be hard to read or electrically noisy; moreover, even an instrument in perfect working order will have a built-in uncertainty due to its design. The thing being measured may not be stable, for example a melting ice cube or an erupting volcano. The measurement process may be difficult, involving, for instance, uncooperative small animals or the subjective judgement of the intensity of a sound. In fact measurements will always depend to some extent on the individual skills of different operators in setting up an instrument and taking readings, such as their reaction times in using a stopwatch. Additionally, measurements must be representative of the thing being measured: do not measure urban air quality on an unusually breezy day, or production-line quality on a Monday morning. Finally, the environment may affect a measurement in overlooked ways, such as the expansion of a steel rule on a very hot day.

Careful measurers must avoid introducing avoidable uncertainties and also estimate the size of unavoidable uncertainties. This estimate requires statistical theory, but even without it we can still get a clear idea of the measurement problem. Regarding the tape measure, has it been calibrated, and if so, what is the uncertainty? When was it calibrated? Could it have stretched or got shorter through bending in the meantime? And what is its resolution (its smallest divisions)? Regarding the string, does it lie straight and if not, how can any bends or kinks be allowed for? If it has been stretched, is it over-stretched? Are its ends frayed or well defined? How do temperature and humidity affect its length? As for the measuring process, how well can the measurer line up the beginning of the string with the start of the tape measure? By what method can the string and the tape measure be consistently aligned? And how repeatable are the measurements?

Naturally, no string measurer will bother to repeat measurements over and over again, but scientists certainly will, in order to check a surprising or controversial result. As Philip Anderson, a Nobel laureate in physics, once observed: 'Many good scientists instinctively distrust a measurement which is *always* on the ragged edge of "statistical significance", and have learned to be very sceptical of marginal statistics.'

# Telescopes

The early telescopes were refracting instruments, which focused light into the eye using lenses. Reflecting telescopes, by contrast, where light is collected and focused by a curved mirror, detected and then magnified by an eyepiece, came a little later, invented by Newton in 1668; in 1781, William Herschel used one to discover the planet Uranus; gradually they dominated astronomy. The 'size' of such a telescope is the diameter of its primary mirror. Double the size and one quadruples the sensitivity: the telescope can see celestial bodies one quarter as bright or, if one prefers, a body of a given brightness twice as distant.

An approximate doubling occurred over three generations of famous telescopes in the United States, from the 2.5-metre Hooker telescope on Mount Wilson (1917) to the 5-metre Hale telescope on Mount Palomar (1948) to the twin 10-metre Keck telescopes on Mauna Kea, Hawaii (1993).

Simultaneously, the detector improved from photographic plates that registered only a few per cent of the light falling on them to electronic detectors that are almost 100 per cent efficient. In addition, computers are increasingly able to compensate for the distortion caused by atmospheric turbulence.

Each Keck telescope weighs 300 tons and is 8 stories tall. Their mirrors are mosaics of 36 hexagonal segments, 1.8 metres across, the hexagon allowing the creation of a hyperboloidal reflecting surface. The segments' alignment must be exceptionally precise, not only to minimize the effect of the joints but also to withstand buffeting on a wind-scoured mountainside.

There are plans to build 24- and 30-metre telescopes and, amazingly, a 100-metre

Above: **Galileo Galilei (1564–1642), one of the founders of modern science and a pioneer of the refracting telescope. In 1610, he squinted through a lens to observe Jupiter's four moons, and recorded tables showing the times of their appearance and disappearance around Jupiter. Galileo suggested the tables could be used by navigators as a celestial time check to calculate their longitude.**

Left: **Keck I and II telescopes on Mauna Kea, Hawaii.**

telescope named Owl (for its night vision but also for '*Over*whelmingly *l*arge'). Owl's principal investigator, Robert Gilmozzi, comments: 'For visible and near-infrared light, ground-based telescopes offer higher resolution and sensitivity at lower cost than orbital observatories' (such as the Hubble space telescope). With anti-turbulence software, the planned giant telescopes should be able to see the earliest stars to be born, planets around other stars, and possibly even sister planets of Earth.

Left: **At the Mount Wilson observatory in California, in 1931, Albert Einstein was shown the photographs of distant galaxies taken with the 2.5-metre reflecting telescope by Edwin Hubble and his assistant Milton Humason. Einstein accepted that the images proved the expansion of the Universe and immediately told waiting journalists that he would abandon the static model of the Universe he had favoured while developing relativity. The photograph shows Einstein with Charles E. St John.**

Below left: **Airy's transit circle, a refracting telescope built at Greenwich in 1850 and named after the astronomer royal of the time, George Biddell Airy. 'The great telescope was placed exactly on the meridian rotating upon precision trunnions. Declinations were read off against a 6-foot-diameter graduated circle. The man on the extreme right reads the scale with 6 micrometer microscopes placed at 60 ° apart, set into the stone pier' (Allan Chapman in *Dividing the Circle*). In 1884, the instrument came to designate the International Greenwich Meridian as the standard for Greenwich Mean Time; it was still in use a century later; and it remains in full working order.**

# Microscopes

In 1679, one of the pioneers of microscopy, Antoni Van Leeuwenhoek, communicated the fact to the Royal Society in London that the number of 'little animals' (spermatozoa) he had just seen in the milt of a cod – 150 billion – was far greater than the total number of people that the planet could support. For two centuries after this, especially in the 19th century, the magnifying power and resolution of microscopes improved, as lenses were freed of chromatic and spherical aberration. But in 1896, Lord Rayleigh showed that the resolution limit for an optical system was determined by the wavelength of light. Nothing smaller than this wavelength could be viewed because of the effect of diffraction in the aperture of the microscope: atoms would remain invisible.

The first images of atoms were obtained in 1955, and by the 1990s peering at atoms was routine. But Rayleigh was correct in his analysis: instead of visible light, a beam of electrons is used to probe a surface. According to quantum mechanics (see pp. 83–4), electrons can behave like waves as well as particles, but their wavelength is much shorter than that of visible light and hence they can probe matter at a smaller scale than light, at the level of atoms.

An electron beam, focused by electrical or magnetic fields, is used in the scanning transmission electron microscope. In the more advanced scanning tunnelling microscope (STM), based on a quantum-

Above: **The frontispiece of Robert Hooke's *Micrographia*, 1664. Hooke was then curator of experiments at the Royal Society, which published his book in a folio, with wonderfully detailed, fold-out, copperplate engravings, some by Christopher Wren. It caused a sensation and became a bestseller. Samuel Pepys bought it immediately and sat up till 2 a.m. reading it; he noted in his diary that it was 'the most ingenious book I ever read in my life'.**

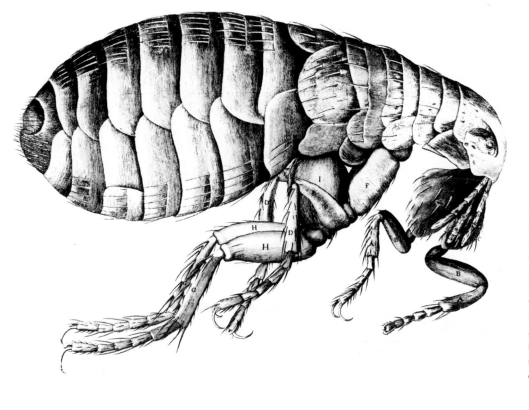

Left: **Hooke's newly revealed invisible world, magnified from 50 to 100 times, showed needles, soot, flies and fleas (seen here), linen, mould, cork, feathers and more, including a plant cell – a term coined by Hooke because its shape reminded him of the cell of a monk.**

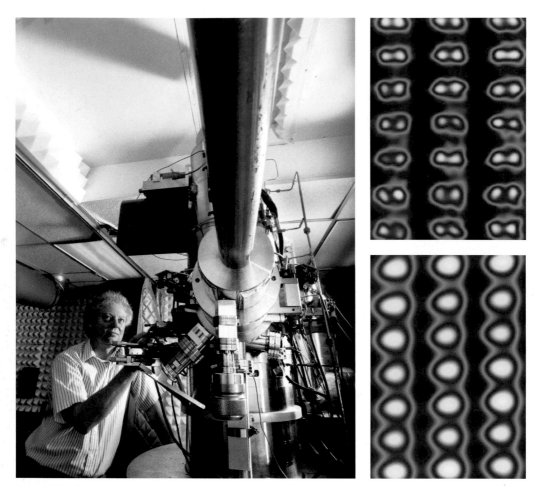

**Atoms under the microscope.** Far left: **The aberration-corrected scanning transmission electron microscope (Stem) can view atoms in a silicon crystal.** Above left: **when the crystal is viewed from a particular direction, the silicon atoms line up in pairs of columns that are a mere 78 picometres ($10^{-12}$ m) apart.** Below left: **The second image simulates what would happen if the aberration correctors were removed: the resolution would be worse and the pairs would merge into single, elongated peaks. Stem images are powerful aids in understanding the macroscopic properties of materials.**

mechanical phenomenon known as 'tunnelling', there is no beam, instead electrons are induced by a voltage to tunnel between the sharpened tip of the STM and the surface under examination, creating a varying electric current that can be translated into a topographical map of the surface. However the STM requires the surface to be an electrical conductor, and many surfaces of interest are non-conductors. In the most recent advance, a mechanical rather than electronic technique is employed. The atomic force microscope (AFM) has a scanning tip with a thin cantilever attached to it. The AFM works by measuring the tiny forces between the scanning tip and the sample through the deflection of the cantilever – in principle, the tip does not have to touch the sample at all. For biologists, the AFM is excellent for imaging protein molecules without disturbing them, though it cannot see the thousands of atoms inside proteins. For materials scientists, the AFM has been more powerful, indeed revolutionary, fuelling the explosion of interest in nanotechnology.

# Thermometers

Almost 300 years have passed since the physicist Daniel Gabriel Fahrenheit constructed the first accurate thermometer, using mercury in glass, in 1714. But surprisingly it was not introduced into medicine until the mid-19th century. Physicians were conservative and treated fever more often as a syndrome or disease than as a symptom. They had long recognized some relation between an elevated body temperature and disease but were yet to grasp fully the equivalence of a 'high temperature' and a fever. In the United States, the leading advocate of thermometry, the physician Edouard Seguin, faced apathy from the medical profession. In 1871, he pleaded with his colleagues: 'I think it is our duty to teach [each mother]… not only the use, but the philosophy of thermometry.' Then, when

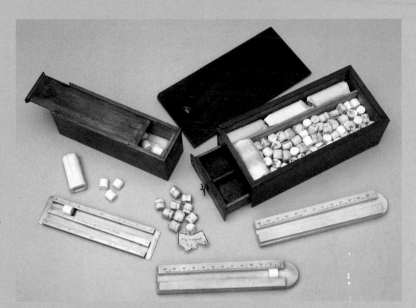

'Neighbours, quacks and mediums proffer in vain their nostrums, she stands by her thermometer, knowing that a calm record of a day's fever brings more hope than a dishevelled therapeusis.'

Various temperature scales have been used since the 18th century – for example, the Réaumur scale to measure the winter temperature during Napoleon's military retreat from Russia in 1812–13 (see p. 13) – but the best known are the Fahrenheit and the Celsius scale. In addition, the Système International and the majority of scientists use the Kelvin scale of absolute temperature.

Fahrenheit borrowed from an earlier researcher, Ole Roemer, the concept of fixed points to define his mercury-in-glass scale. Roemer had used an ice–salt mixture to fix his minimum temperature and the boiling point of water his maximum, covering a scale of 60 degrees, a figure apparently chosen for its use in measuring angles and time.

Above: **Pyrometer presented to King George III by the potter Josiah Wedgwood in the 1780s. Wedgwood needed to know the temperatures in his kilns, which far exceeded the boiling point of mercury (357 °C). His pyrometer relied on measuring the shrinkage of clay when fired, which persisted on cooling and depended only on the highest temperature attained, not on the length of time the clay spent in the kiln. Using specially made clay, Wedgwood drew up a temperature scale in degrees Wedgwood (0–160) with a maximum at over 20,000 °F, though this was undoubtedly an overestimate.**

Left: **Mercury thermometers under test in a furnace, early 20th century.**

(Roemer was an astronomer, who measured the speed of light in 1676.) Fahrenheit decided to expand Roemer's scale by quadrupling each degree to cover a total of 240 degrees. He fixed the temperature of a mixture of ice, salt and sal-ammoniac at 0 °F, the freezing/melting point of ice at 30 °F, and the normal temperature of the human body (blood heat) at 90 °F. But after some experimentation, Fahrenheit revised the scale: the zero remained unchanged but he moved the ice point up to 32 °F, while body heat moved to 96 °F and water's boiling point came down from 240 °F to 212 °F. The scale and its evolution were not particularly logical, nevertheless Fahrenheit's degrees were widely accepted since his thermometers worked well, partly because he had devised a method of purifying mercury by filtering it through leather so that impurities did not block the thermometer's thin glass tube. We still use Fahrenheit's figures for the ice and boiling points (at standard atmospheric pressure), though not his figure for blood temperature, which now falls in the range 98.4–98.6 °F for a healthy person.

However, the new scale quickly had competition. Not long after Fahrenheit's death in 1736, Anders Celsius put forward a scale of 100 degrees with two fixed points: the freezing point and the boiling point of water. Strangely, Celsius chose 100 °C as the freezing point and 0 °C as the boiling point. But after one of his pupils, following Celsius's premature death, reversed his fixed points, the 100-degree scale began to catch on and became known as the Celsius scale in the first half of the 19th century. Yet many people imagined then, and still do, that 'degrees C' stands for 'degrees Centigrade',

because there are 100 degrees in the familiar part of the scale. Only in 1948 did the Ninth General Conference on Weights and Measures make the change to *Celsius* official.

The idea of absolute zero and absolute temperature was conceived in 1848, before refrigeration had been invented. Lord Kelvin proposed it simply from his observations of what happened to the volume of a gas as its temperature fell. For every 1 °C fall in temperature, the gas contracted almost uniformly by 1/273th of its volume at 0 °C. So Kelvin thought that the volume must fall to zero, 'absolute zero', at −273 °C. Although we now know that matter cannot physically disappear in this way – and indeed it is theoretically impossible for us to reach absolute zero – Kelvin's extrapolation was very close to the modern value for the lowest temperature.

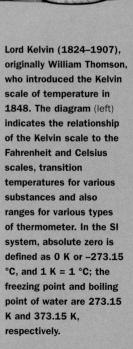

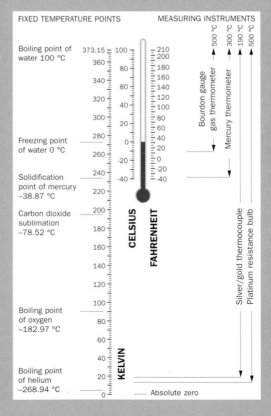

Lord Kelvin (1824–1907), originally William Thomson, who introduced the Kelvin scale of temperature in 1848. The diagram (left) indicates the relationship of the Kelvin scale to the Fahrenheit and Celsius scales, transition temperatures for various substances and also ranges for various types of thermometer. In the SI system, absolute zero is defined as 0 K or −273.15 °C, and 1 K = 1 °C; the freezing point and boiling point of water are 273.15 K and 373.15 K, respectively.

# Barometers

The concept of the barometer presumes the existence of atmospheric pressure, which presumes the possibility of a vacuum: a void from which all air has been removed. Evangelista Torricelli invented the barometer in 1644 following an earlier suggestion from his mentor Galileo. Torricelli filled a glass tube, 120 centimetres long and sealed at one end, with mercury, and then inverted the tube in a dish. Instead of all the mercury staying in the tube, a column about 76 cm tall remained, with a space above it. Torricelli realized that atmospheric pressure, acting on the mercury in the dish, was supporting the column, and the space had to be a vacuum. Over a period of time, he observed that the height of the mercury varied a small amount from day to day with varying atmospheric pressure – the basic principle of the barometer. Torricelli concluded that 'the quicksilver rises to the point at which it comes into equilibrium with the weight of the outside air pressing down on it', and that all of us 'live submerged at the bottom of an ocean of the element air'.

The commonest pressure units in modern barometry are millimetres of mercury (mmHg) or millibars (mb), although the SI unit is the newton per square metre, named the pascal (Pa). Normal atmospheric pressure at sea level is 760 mmHg/1013.2 mb (101,320 Pa), or in imperial units 14.7 pounds per square inch, that is half the normal pressure in a car tyre. (However, a tyre gauge measures from ambient, i.e. atmospheric, pressure, not from zero pressure.) Of course atmospheric pressure falls with altitude, as discovered by Blaise Pascal soon after the experiments of Torricelli; at the top of Mount Everest, it is only about 300 mb. But a climber who managed to get a barometer to the summit would also have to adjust the reading for temperature and local gravity, both of which affect the mercury in barometers.

One other type of barometer is commonly used, especially in aircraft altimeters: the ancroid barometer, invented in 1844 by Lucien Vidie. Its name derives from the Greek for 'not wet'. The aneroid employs no liquid, only an evacuated capsule with a flexible diaphragm that deflects with change in pressure and is coupled via a spring and lever to an indicating needle. It is much more portable and more convenient, especially for automated recording, than the mercury barometer, but not as accurate.

Top: **Evangelista Torricelli (1608–47), inventor of the barometer. He was Galileo's final assistant.**

Above: **Robert Fitzroy (1805–65), first director of the Meteorological Department, in the uniform of a British vice-admiral.**

Left: **'Fitzroy' mercury barometer, a design first introduced in the 1860s.**

# Seismographs

The earliest known seismograph was the work of the Chinese astronomer and mathematician Chang Heng: it consisted of eight dragons mounted around a base consisting of eight toads with open mouths; even a slight tremor would shift a pendulum within the base, activating lever devices that caused a bronze ball to drop from a dragon's mouth into a toad's mouth with a resonant clang.

An Englishman in Japan, John Milne, invented the modern seismograph in the 1880s. It employed the pendulum principle too, as indeed do all seismographs. The inertia of a large, freely suspended weight makes it lag behind the movement of its frame: the mechanism of the seismograph arranges for this relative motion to be recorded. By using three pendulums that can bob up and down and swing from side to side like a door on its hinges, both the vertical and the twin horizontal vibrations of an earthquake can be monitored.

Milne's first version used a stylus attached to the weight that inscribed a trace on revolving smoked paper – a method sometimes still employed in the field for its reliability. Afterwards, in 1893, Milne adapted the instrument to record the trace on revolving photographic film. Nowadays, the relative motion is converted into an electrical signal, amplified electronically thousands or even hundreds of thousands of times, and recorded by a stylus on paper, by magnetic tape or by computer.

There are two basic types of earthquake wave (below left): body waves, that are transmitted from the earthquake focus underground to the nearest point on the surface (the epicentre); and surface waves, that are produced in the ground by the transformation of some body waves once they reach the surface. The body waves consist of P (primary waves), and S (secondary) waves, while the two types of surface wave (below right) are named after the mathematician A. E. H. Love and the physicist Lord Rayleigh who defined them.

Far left: **A section of the earliest known seismograph, designed by Chang Heng in AD 132.**

Below: **The essentials of the modern seismograph.**

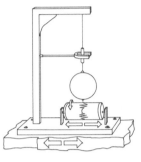

Horizontal Earth motion

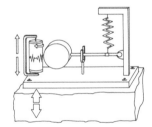

Vertical Earth motion

Left: **Waves in the ground. P waves travel fastest (up to 6.5 km/s); the first movement felt in an earthquake is therefore caused by a P wave. It moves fastest because it is longitudinal, like a sound wave. The S wave, by contrast, is transverse: it propagates with a side-to-side shearing motion, like a radio wave. It makes the ground move both vertically and horizontally.**

## Body waves

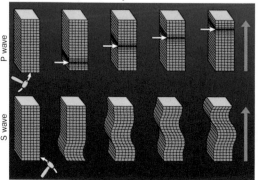

P wave

S wave

## Surface waves

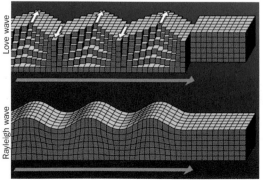

Love wave

Rayleigh wave

# Geiger Counters

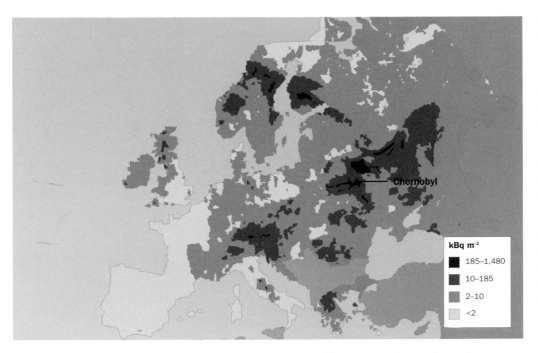

In a typical residential location, ambient radioactivity will make a sensitive Geiger counter click at an average rate of about once every 2 seconds. After rainfall, which brings down dissolved radioactive substances, the rate may rise to almost once a second. This means 1 nuclear disintegration per second, or 1 becquerel. Such a low figure for background radiation indicates just how serious was the nuclear accident at Chernobyl in 1986.

The prototype of the Geiger counter was invented in 1908 in Manchester, while Hans Geiger was working with Ernest Rutherford. It was intended to count alpha particles from radioactive decay – work that led, in 1911, to Rutherford's shattering discovery of the atomic nucleus. In 1928, now back in Germany, Geiger improved the design with his student Walther Müller. The basics consist of a tube containing a low-density gas, usually argon, and two electrodes, positive and negative. The voltage across them is slightly less than required to produce a discharge – a bit like a flickering fluorescent bulb. Radioactive particles entering via a window, being highly energetic, initiate a discharge by knocking electrons off the gas atoms, *ionizing* them (see p. 89). The electrons flow towards the positive electrode and the positively charged ions towards the negative electrode, colliding with other gas atoms and ionizing them too, thereby creating an avalanche of electrons intense enough to create a pulse of electric current detectable after amplification as a click in a loudspeaker. One alpha (or beta) particle, one click. Counting rates of up to a few hundred per second are practicable – alternatively the current can simply be measured in amperes.

Left: **This map shows the levels of radioactivity from the world's worst nuclear accident, at Chernobyl in 1986, during which 6.7 tonnes of radioactive material spread for hundreds of kilometres around the site. The map refers to caesium-137, the most prevalent isotope, which has a half-life of 30 years (see p. 86) – in other words its radioactivity will have fallen to half its 1986 level by 2016. While this means that much of the abandoned area around the reactor should become habitable in the coming decades, the accident has already caused about 4,000 cases of thyroid cancer, mainly in children and adolescents, according to a report by the United Nations and the governments of Ukraine, Belarus and Russia, released in 2005.**

Above: **Hans Geiger (1882–1945), who invented the Geiger-Müller counter for measuring radioactivity. It is suitable for detecting alpha and beta particles, but is not reliable for gamma rays.**

# Spectrometers

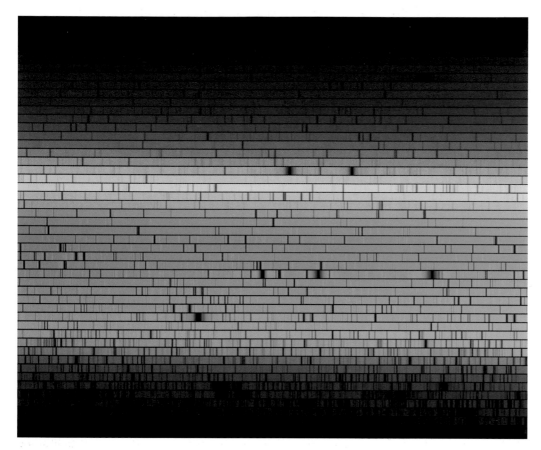

Until the introduction of spectroscopy in the 19th century, astronomers could learn about the chemical and physical conditions of celestial bodies only by chemically analysing meteorites that fell to Earth. Nowadays, by analysing the differing wavelengths of light emitted by stars with spectrometers, they are able to gain deep insight into stellar temperatures and the structures and proportions of the chemical elements, atoms and molecules present in the far reaches of the Universe. For example, spectral analysis shows that the atmosphere of Venus is 96.5 per cent carbon dioxide, and the outer region of the Sun is about 90 per cent hydrogen.

Any experiment in spectroscopy involves: a source of radiation (whether visible or otherwise); a refractive prism or diffraction grating that splits the radiation into its constituent wavelengths; detectors for observing or recording details of the spectrum; measurements of wavelengths and intensities; and interpretation of the data by comparison with spectral data from terrestrial physics and chemistry. Instruments designed for direct visual observation are called spectroscopes, while those that use photography or other means of recording data are known as spectrographs.

Absorption lines in a spectrogram of visible light reaching Earth from the Sun. The black lines – indicating no transmitted light – are caused by various wavelengths that have been absorbed by chemical elements in the Sun. The opposite process, in which high-energy atoms lose energy and emit radiation, produces an emission spectrum with bright lines. The absorbed energy is used in the solar reactions that keep stellar fires burning, such as fusion between hydrogen nuclei to form helium. Each chemical element has a spectral fingerprint. Hydrogen has both an absorption and an emission spectrum, which consists of five series of lines: one series in the visible (included in the photograph): one in the ultraviolet and three in the infrared. In 1913, Niels Bohr took the first step in explaining these series theoretically by relating the energy changes represented in the lines to the existence of quantized energy levels in the hydrogen atom (see pp. 83–4).

# Lasers

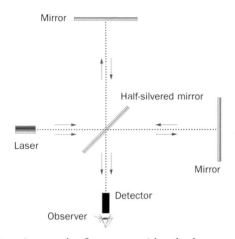

Mirror

Half-silvered mirror

Laser

Mirror

Detector

Observer

Einstein was the first to consider the key concept of stimulated emission of light – the word laser stands for 'light amplification by stimulated emission of radiation' – in 1917. Building on his revolutionary quantum theory, which treated light as both a particle and a wave, he predicted that one quantum of light (a *photon*, analogous to an electron) could stimulate an atom in a high-energy state to emit two photons of the same energy – thus amplifying light.

But Einstein did not propose a second key concept, the *coherence* of light. Coherence means that the two stimulated photons are,

so to speak, identical copies. Not only do they have the same frequency (since their energy is the same) when behaving as waves, they are in step too. In other words, their peaks and troughs are similarly spaced, with a fixed phase relationship, and so they combine, or 'interfere constructively', and reinforce each other. A coherent beam of light has waves that are in phase; an incoherent beam – that is, practically all familiar forms of light (including sunlight) – has peaks and troughs not aligned in the same way, which tend to cancel each other out, or 'interfere destructively'. Its single frequency and coherence is what gives a laser beam its unique focus and power.

Far left: **For length measurement, the laser interferometer is the most accurate instrument available. In the interferometer diagram, a laser beam is split into two beams, at right angles to each other, by a half-silvered mirror. The two beams are then reflected by two adjustable mirrors in the 'arms' of the interferometer, recombined at the half-silvered mirror, and reflected into the observer's detector. If the beams travel precisely the same distance, they arrive precisely in phase and interfere constructively; but if one beam travels half a wavelength (about 300 nanometres) further than the other, they arrive out of phase and interfere destructively, creating a pattern of light and dark interference 'fringes' observable by the detector. Thus, a minute difference in length between the arms – about a nanometre – can be measured.**

The origin of the laser. Many scientists, including Einstein, contributed to the laser. Gordon Gould named it in his legally notarized notebook of 1957 (left). Alexander Prokhorov, Charles Townes and Nikolai Basov (above left) shared the Nobel prize for its invention, in 1964. The first working laser was built by Theodore Maiman in 1960.

# National Institutions for Measurement

Left: **Views of the Greenwich observatory, as seen in 1885, the year after the international acceptance of Greenwich Mean Time. On the far left is Airy's transit circle (see p. 68) and in the centre the clock observed by the 'time carrier' Ruth Belville (see p. 62).**

Astronomy created what might be considered the first national institutions for measurement: the Paris observatory, founded by King Louis XIV in 1667, and the Greenwich observatory, founded by King Charles II in 1675. But their functions were limited to astronomy, navigation and surveying. London's Royal Institution (1799), had a broader remit, covering all that fell within 'the application of science to the common purposes of life', but it did not attempt to establish national standards in weights and measures. Not until the last quarter of the 19th century were recognizable institutions of national standards established: the Bureau International des Poids et Mesures in France (1875), the Physikalisch Technische Reichsanstalt in Germany (1887), the National Physical Laboratory (NPL) in Britain (1900) and the National Bureau of Standards in the United States (1901). The reason for this clustering was of course the rapidly increasing sophistication of technology and the rapidly advancing requirement for precision and accuracy in science. Industry and science together were demanding standardization of physical quantities such as time, length, mass, force, pressure and capacity.

In Britain, from the beginning, the government-funded NPL was managed by a body consisting of fellows of the Royal Society and representatives from industry. Inevitably, there was conflict between scientific research and commercial activity at NPL. The government's civil servants were unsympathetic to fundamental science, and were convinced that scientific research should in the long run be financially self-supporting. In 1965, the governing body was

Below: **Lord Rayleigh (1842–1919), originally John William Strutt, the physicist who became the first chairman of the governing body of the National Physical Laboratory in 1899. He was awarded the Nobel prize in 1904, and became president of the Royal Society in 1905.**

disbanded, the link with the scientific community formally broken, and the NPL absorbed into the Ministry of Technology. Then in the 1990s, the government disgorged the NPL by withdrawing direct government support and turning it into more of a business concern than a government department; however, the link with the Royal Society was restored.

But although from the outset much of the NPL's work was fee paying and humdrum – testing thermometers, standardizing and calibrating scientific instruments sent for testing, and undertaking various physical and chemical analyses – after the First World War the laboratory did establish a reputation for research in many fields, which increased after the Second World War. Some pioneering research on computing came out of NPL in the 1940s and 1950s, notably the Pilot Automatic Computing Engine (Ace) designed by Alan Turing; as did the earliest reliable atomic clocks in the 1950s; and the concept of computer networks and packet-switching in the 1960s, which contributed to the invention of the internet in the United States.

A history of the NPL, *A Century of Measurement* by Eileen Magnello, gives a good idea of its range of activity at mid-century. During 1950, she notes, advice was sought from the NPL on: the acoustics of a public address system in St Paul's Cathedral; the acoustics of Winchester Cathedral; the evaporation of trilene in a device suggested for analgesia in childbirth; the temperature distribution within milk bottles during sterilization; finding a possible method of

locating leaks in buried water pipes; the thermal and humidity problems involved in the subterranean growing of mushrooms; and the precautions to be taken when a radium needle burst open during sterilization at a hospital.

Technical challenges, by their very nature, change a lot over the decades; but the need for standardization and national standards institutions remains. In the early 21st century, much calibration work has moved online. The internet allows major national standards institutions to perform and request some calibrations remotely with minimal physical transportation of artifacts or equipment. With no more than a standard connection (the same as that used in a browser), software at the institution can control the measuring equipment located at the distant laboratory, analyse the measured data, and issue a certificate.

Below: **Optical glass manufacture at the National Bureau of Standards, 1928. The 1.8-metre glass disc weighing 1.7 tonnes – at the time the largest in the United States – was for the mirror of a reflecting telescope. Four discs were poured and cracked during cooling, but the fifth one, because it was cooled very slowly, survived. This success of the government-run National Bureau influenced the casting of a 5-m mirror for the Hale telescope on Mount Palomar by a private US company in the 1940s. Since 1988, the National Bureau of Standards has been known as the National Institute of Standards and Technology.**

# II MEASURING NATURE

To speak of stars, galaxies and the Big Bang origin of the Universe, and of electrons and quarks and even string theory, all in one breath, is not difficult. To write down their relative sizes is manageable too, using logarithmic notation or lots of zeroes. But to grasp what such comparisons mean is nearly impossible. The mind struggles enough with great natural forces like hurricanes and tsunamis. Galaxies are vastly bigger than the Earth–Sun distance, and strings, if they exist, are incredibly smaller than the size of a bacterium – a range of magnitude we can just about conceive.

Even scientists sometimes take refuge in non-standard units, more 'human' than, say, gigametres ($10^9$ m) and nanometres ($10^{-9}$ m). Astronomers like the 'astronomical unit' (AU), equalling the mean distance between Sun and Earth; from the Sun to Jupiter is 5.20 AU, a figure easier to remember than the metric distance, 778 gigametres. Chemists are fond of the angstrom (Å), 0.1 nanometre, for measuring molecular distances; the radius of the chlorine molecule is about 1 Å.

The instruments that probe galaxies, volcanoes and atoms – from giant telescopes to satellites and lasers – are triumphs of human ingenuity. They convert these natural phenomena into numbers. How we humans choose to interpret their measurements is science's continuing intellectual journey.

The furthest and oldest reaches of the visible Universe, as seen by the Hubble space telescope in an Ultra Deep Field image, 2004. This humbling image shows the first galaxies to emerge after the Big Bang, when the Universe was just 400 million years old. The latest estimate for the age of the Universe is around 13.7 thousand million years

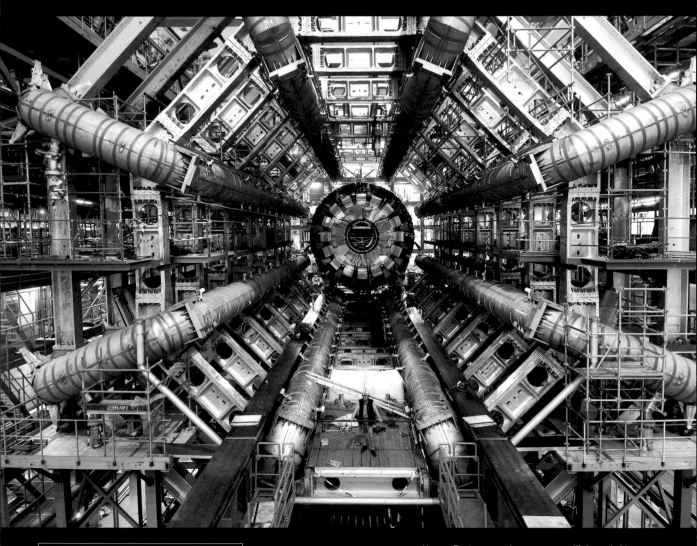

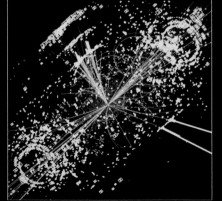

Above: **Proton smasher. The huge detectors of the world's largest particle accelerator, the Large Hadron Collider, are designed to measure the fallout from the collision of protons accelerated to speeds much higher than possible in earlier accelerators. Physicists hope that in the explosion of subatomic particles that sprays out from the collisions** (left)**, some new and exotic ones may be discovered. The Large Hadron Collider is located underground near Geneva at the European Council for Nuclear Research (Cern) on the border of France and Switzerland. It has been funded and built in collaboration with over 2,000 physicists from 34 countries, universities and laboratories.**

# Atoms and Quantum Theory

Is there any theoretical limit to how far we can chop up matter? Do we eventually reach indivisible particles? Is nature fundamentally continuous and wave-like, or fundamentally discontinuous and quantized?

This debate began in the 5th century BC with the Greeks' atomistic theory, revived during the scientific revolution of the 17th and 18th centuries, intensified with quantum theory in the 20th century, and continues in the present day with its opposed concepts of analogue and digital and the wave-particle 'duality' of entities such as the electron. The philosopher Bertrand Russell is supposed to have asked: Is the world a bucket of molasses or a pail of sand? In mathematical terms, 'Is the world to be described geometrically as endless unbroken lines, or is it to be counted with the algebra of discrete numbers? Which best describes nature – geometry or algebra?', asks a contemporary physicist, John Rigden.

During the 19th century, evidence for atoms and molecules accumulated. The chemist John Dalton defined an atom as the smallest part of a substance that could participate in a chemical reaction. Dalton argued that chemical elements such as oxygen and iron combined in definite proportions, and he produced the first table of comparative atomic weights. In the 1860s, the periodic table was developed by Dimitri Mendeleev, in which the elements were classified according to their atomic weight in groups with similar properties. (Today, the periodic table is based on atomic number, not weight – see p. 88.) Later, physicists were able to explain phenomena such as heat and

temperature statistically, in terms of enormous numbers of moving or vibrating atoms and molecules, by assuming that these particles had kinetic energy, which could increase or decrease as they moved faster or slower. But as yet there was no overwhelming experimental proof that such invisible microscopic entities existed.

Then in 1905 Einstein showed how the kinetic theory of atoms and molecules could explain the puzzling phenomenon of Brownian movement. In 1827, the botanist Robert Brown had reported the erratic movement of very fine pollen particles suspended in water. By Einstein's time, the phenomenon had been generalized to include erratic movement wherever small inanimate particles – for instance, finely ground glass – were suspended in liquids or gases. Therefore the cause had nothing to do with botany or living things. Einstein proposed that these tiny particles were being agitated by collisions with constantly moving liquid and gaseous molecules.

Eminent physicists were initially unpersuaded. For one thing, atoms and molecules were surely far too small to affect a comparatively huge pollen particle; the relative size of a molecule and a pollen

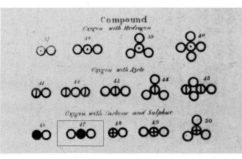

⊙ Hydrogen
◗ Nitrogen
● Carbon
○ Oxygen
⊕ Sulphur
⊗ Phosphorus
⊙ Alumina
◐ Soda
◑ Potash
Ⓒ Copper
Ⓛ Lead
Ⓘ Iron

**Dalton's atoms and molecules. In 1808, John Dalton published his *New System of Chemical Philosophy*, with a newly devised system of chemical symbols. Some of his own drawings of the combinations of oxygen are shown** at left **(including carbon dioxide, which has been highlighted), and modern versions of his elemental symbols** above. **Many molecular formulae were incorrect, but Dalton's concept and his list of atomic weights were influential.**

particle was much smaller than, say, the relative size of a mosquito and an elephant. For another, the moving molecules presumably 'pin pricked' a pollen particle from all directions, creating an average effect of zero. Einstein, however, demonstrated that the observed zigzagging of the pollen particles could arise from 'crowd behaviour' by atoms and molecules. In other words, localized statistical fluctuations of large groups of atoms and molecules temporarily moving en masse first in one particular direction and then in another, and at the same time pushing a pollen particle this way and that. From theory, Einstein calculated that particles in water at 17 °C with a diameter of a thousandth of a millimetre – that is, 10,000 times bigger than atoms – should move a mean horizontal distance of 6 thousandths of a millimetre in 1 minute. Jean Perrin soon confirmed this prediction in the laboratory.

Quantum theory was born in 1900 with the work of Max Planck. He considered the energy of the heat emerging from a glowing cavity. He tried to devise a theory to explain how the measured energy varied over different wavelengths and at different temperatures. But when he treated the heat as a continuous wave, his wave model did not agree with experiment. Only when he assumed discrete values for the energies of the 'resonators' (atoms) in the walls of the cavity that were absorbing and emitting heat, did theory match experiment. Instead of continuous absorption and emission of energy, energy was exchanged between heat

and atoms in packets or quanta. Moreover, the size of a quantum was proportional to the frequency of the resonator, which meant that high-frequency quanta carried more energy than low-frequency quanta. Being a believer in nature as a continuum, and an innately conservative man, Planck did not feel at all comfortable with his calculation, but he reluctantly published it.

Einstein was much bolder. He decided that not just the exchange of energy between heat/light and matter was quantized – *light itself was quantized*. In 1905, he stated: 'when a light ray is spreading from a point, the energy is not distributed continuously over ever-increasing spaces, but consists of a finite number of energy quanta that are localized in points in space, move without dividing, and can be absorbed or generated as a whole.' Instead of moving particles, Einstein visualized a light beam as moving packets of energy. In the 1920s, when this revolutionary concept was finally accepted, the packets were named 'photons'.

Quantum theory is an immensely powerful tool. By assuming that the energies of the particles in the atom are quantized, scientists have been able to explain, for example, spectral absorption and emission lines, the different properties of chemical elements, and the spray of subatomic particles from collisions in accelerators. But Planck's discomfort still resonates. 'Science cannot solve the ultimate mystery of nature,' he once wisely remarked. 'And it is because in the last analysis we ourselves are part of the mystery we are trying to solve.'

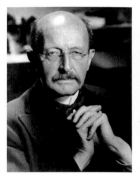

Above: **Max Planck (1858–1947), the physicist who reluctantly introduced the quantum into science. Planck was a strong advocate of Einstein's theory of relativity, but distrusted Einstein's quantum theory.**

Above: **Brownian movement, as observed by Jean Perrin. He recorded the positions of three granules of mastic in liquid solution every 30 seconds and then connected the positions with straight lines to reveal the zigzag movement. The experiment was the first conclusive demonstration of the existence of atoms and molecules.**

# Atomic Clocks

What is a clock? Forget rotating hands or revolving numbers. The fundamental fact about a clock, as already noted, is that it is driven by a cyclic mechanism – an oscillation – of uniform frequency. This is true whether the oscillation is that of a pendulum, a spring-coupled balance, the rotation of the Earth, the moons of Jupiter observed by Galileo, or an atom or molecule that absorbs a burst of energy (i.e. cyclic radiation) of a frequency determined by its electronic structure, during the transition or 'jump'

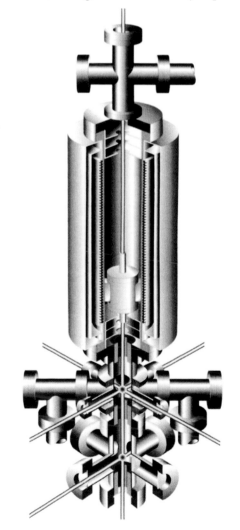

of a subatomic particle such as an electron from a lower-energy quantum level to a higher-energy level.

This last mechanism now fixes the time to many splits of a second. London's world-famous clock Big Ben is, so to speak, just the front man for atomic time. Since 1967, astronomical time measured at Greenwich has officially given way to atomic time. In the words of the UK's National Physical Laboratory, 'The second is the duration of 9,192,631,770 periods of the radiation corresponding to the transition between the two hyperfine levels of the ground state of the caesium atom.'

That may not mean much to most of us, but what is evident to all is that atomic quantum phenomena must provide a standard exactly reproducible anywhere in the world with a suitably equipped laboratory. Atomic structure is universal, so that a caesium atom – unlike a pendulum – will always keep the same time, thousands of years from now or in a distant galaxy.

Such timekeeping may be irrelevant to our everyday sense of time. But it is crucial to the operation of the internet, email, television, the Global Positioning System, the power industry, transport and financial systems. It is vital to the emergency services, train companies, cash machines and mobile phone billing systems. For several thousand years, mankind managed well enough without atomic time, but without it now, modern civilization would quickly come to a halt.

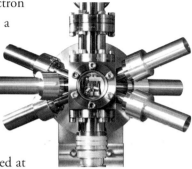

Above: **Next-generation atomic clocks will use ion traps, in which ions are cooled with lasers to close to absolute zero, keeping them stationary, allowing a longer interaction. Ion traps may have accuracies around 1,000 times higher than current atomic clocks.**

Left: **The caesium fountain is the successor to atomic clocks pioneered by Louis Essen in the 1950s. Four lasers cool caesium atoms by slowing them from 200 metres per second to a mere few centimetres per second. Then, clouds of the cold atoms are launched upwards repeatedly. As they fall back under gravity, they pass through a microwave cavity. Microwave radiation (see p. 98) stimulates the atomic transition that defines the second. By measuring the frequency absorbed, we can measure the passage of time. The longer the interaction with the microwaves, the greater the accuracy.**

# Radioactivity and Isotopes

Around the turn of the 19th century, just as the atomic nature of matter was finally established beyond doubt, it was suddenly made obvious that atoms were not indivisible: a chemical element could be unstable and decay into another element with the emission of radiation. Radioactivity was accidentally discovered by Henri Becquerel while working with a uranium compound, in 1896. In 1898, Pierre and Marie Curie discovered two other strongly radioactive elements, radium and polonium, both of which were naturally occurring; and the same was true of thorium.

Then another mainstay of atomic theory was challenged. It had been generally assumed that all atoms of a chemical element were the same and, in particular, had the same weight or mass. The decay products of both uranium and thorium were now known

to be lead. But in 1913, the existence of different forms of lead in minerals from different sources was demonstrated. The forms were chemically identical – a mixture could not be separated by chemical means – but they had different atomic weights. Frederick Soddy named these multiple forms *isotopes*, meaning 'same place', because they occupied the same place in the periodic table (see p. 88). In due course, many naturally occurring isotopes of the same element – some radioactive, some not – were isolated.

Left: **Marie Curie (1867–1934), who won two Nobel prizes, the first in physics for the discovery of radioactivity (with Pierre Curie and Henri Becquerel) and the second in chemistry (solo) for the isolation of radium.**

Below: **Naturally occurring radioactive isotopes. Six exist for uranium, though only $^{238}$U is relatively common.**

| Isotope | Half-life (years, unless noted) | Isotope | Half-life (years, unless noted) |
|---------|--------------------------------|---------|--------------------------------|
| $^{3}$H | 12.32 | $^{149}$Sm | $10^{16}$ |
| $^{14}$C | 5,730 | $^{176}$Lu | $3.8 \times 10^{10}$ |
| $^{50}$V | $>3.9 \times 10^{17}$ | $^{187}$Re | $4.2 \times 10^{10}$ |
| $^{87}$Rb | $4.88 \times 10^{10}$ | $^{186}$Os | $2 \times 10^{15}$ |
| $^{90}$Sr | 29 | $^{222}$Rn | 3.823 days |
| $^{115}$In | $4.4 \times 10^{14}$ | $^{226}$Ra | 1,599 |
| $^{123}$Te | $1.3 \times 10^{13}$ | $^{230}$Th | $7.54 \times 10^{4}$ |
| $^{130}$Te | $2.5 \times 10^{21}$ | $^{232}$Th | $1.4 \times 10^{10}$ |
| $^{131}$I | 8,040 days | $^{232}$U | 68.9 |
| $^{137}$Cs | 30.13 | $^{234}$U | $2.45 \times 10^{5}$ |
| $^{138}$La | $1.06 \times 10^{11}$ | $^{235}$U | $7.04 \times 10^{8}$ |
| $^{144}$Nd | $2.1 \times 10^{15}$ | $^{236}$U | $2.34 \times 10^{7}$ |
| $^{147}$Sm | $1.06 \times 10^{11}$ | $^{237}$U | 6.75 days |
| $^{148}$Sm | $7 \times 10^{15}$ | $^{238}$U | $4.46 \times 10^{9}$ |

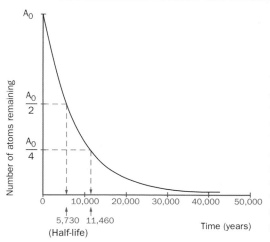

Left: **Radioactive decay. The decay is not linear but exponential as shown in the diagram for naturally occurring carbon-14 ($^{14}$C). This isotope's half-life is 5,730 years, a period that roughly covers the duration of civilization – hence the immense usefulness of carbon dating in archaeology.**

| Dose | Immediate symptoms | Latent phase | Full-blown illness | Prognosis |
|---|---|---|---|---|
| 0.1–0.5 | Mild nausea. | Days to weeks | Slight decrease in blood cell counts. | Almost certain survival. |
| 1.0–2.0 | Mild to moderate nausea and possible vomiting within 3 to 6 hours after irradiation and lasting for up to a day. | 10–14 days | Loss of appetite, fatigue, lowered blood cell counts. | Survival rate 90 per cent. |
| 2.0–3.5 | Nausea and possible vomiting starting 1 to 6 hours after irradiation and lasting for up to 2 days. | 7–14 days | Hair loss, moderate to severe bone marrow damage, severe risk of infection. | Fatality rate 35–40 per cent. |
| 3.5–5.5 | Nausea and vomiting within half an hour, lasting for up to 2 days. | 7–14 days | Hair loss, internal bleeding, severe bone marrow damage with high risk of bleeding and infection, slight gastrointestinal damage. | Fatality rate 50 per cent within 6 weeks. |
| 5.5–7.5 | Severe nausea and vomiting within half an hour, lasting for up to 2 days. | 5–10 days | Hair loss, internal bleeding, severe bone marrow damage leading to complete failure of blood system, high risk of infection, moderate gastrointestinal damage. | Death probable within 3 weeks. |
| 7.5–10 | Excruciating nausea and vomiting within minutes, lasting for several days. | 5–7 days | Severe gastrointestinal and bone marrow damage. | Death almost certain within 3 weeks. Complete recovery impossible. |
| 10–20 | Immediate, excruciating nausea and vomiting and fatigue. | 5–7 days | Severe gastrointestinal, bone marrow and lung damage. Cognitive dysfunction. | Certain death within 5–12 days. |
| 20+ | Coma. | None | Certain death within hours. | Certain death. |

The atomic explanation of isotopes took somewhat longer to establish. First, Ernest Rutherford, who discovered the dense nucleus with its massive, positively charged protons, classified radioactive disintegration into alpha, beta and gamma rays, which he went on to identify as emissions of massive nuclear particles (alpha), low-mass electrons (beta) and high-energy waves (gamma). Then, working under Rutherford, in 1913 Niels Bohr conceived the 'solar system' atomic model of negatively charged electrons orbiting the positively charged nucleus, like planets orbiting the Sun but in orbits fixed by quantum theory. Finally, in 1932, James Chadwick discovered the neutron, a nuclear particle of no charge and about the same mass as the proton.

This new understanding explained isotopes in the following way. A chemical element has an *atomic number*, which corresponds to the number of protons in the nucleus. This number – 92 for uranium – defines an atom as being uranium. But the element also has an atomic weight, which varies according to the number of neutrons in the nucleus. Thus, $^{238}U$ has 92 protons and 146 neutrons in its nucleus (92 + 146 = 238), while $^{235}U$ has 143 neutrons.

Perhaps the most important measure of radioactivity is an isotope's *half-life*. It indicates how long a sample will take to decay into half its original amount. Radioactive half-life is not like human mortality. The half-life does not diminish with time: a sample of carbon-14 ($^{14}C$) will have the same half-life 5,730 years from now, and 5,730 years after that (i.e. after 11,460 years), though of course the sample size will have diminished by half and by three quarters, respectively.

**Deadly radioactive doses. The dose is measured in grays, a unit named after Louis Gray, a radiobiologist who studied radiation-induced cancers.**

# The Periodic Table

The most bewildering aspect of chemistry is the sheer profusion of seemingly unconnected facts about the different elements, their compounds and their reactions. The periodic table provides at least some relief to the hard-pressed memory. But it is much more than that, for it links the atomic structures of the elements to their observed chemical behaviour. It clarifies, for example, why the alkali metals like sodium (Na) and potassium (K) are highly reactive, while other metals, like gold (Au), are comparatively unreactive, and why the elements in the final column, gases like neon (Ne) and argon (Ar), are so unreactive they are called inert or noble gases. As Peter Atkins writes in his metaphorical tour of the table, *The Periodic Kingdom*, 'The kingdom is not an amorphous jumble of regions, but a closely organized state in which the character of one region is close to that of its neighbour.'

Dimitri Mendeleev arranged the first periodic table in 1869 by increasing atomic weight, which he regarded as the only fundamental characteristic of an element. Taking the 61 known elements, he left gaps where he had intuited – apparently in a dream while taking a nap from writing a chemistry textbook – that more elements would in due course be discovered.

The modern periodic table (right) includes well over 100 elements, and the order is now based on atomic number, not weight, starting with hydrogen at atomic number 1 and finishing with the unstable and radioactive elements that begin with thorium at number 90. The atomic number (the number of protons in the nucleus) is also the number of electrons, in an uncharged atom. The electron structure determines an element's basic chemical properties, because it is generally the orbiting electrons, not the dense nucleus, that take part in chemical reactions. The more tightly bound the electrons are to the protons, the less reactive the element, and vice versa. Quantum theory explains the electron structure in detail, but at its simplest we can visualize the electrons filling up 'shells' around the nucleus to a maximum permissible number that depends on the type of shell. Inert gases such as neon (Ne) have a full outer shell, with no space for more electrons, while alkali metals such as sodium (Na) have a single free electron in their outer shell, which can be easily removed (ionized) in reaction. Thus, even though Ne has atomic number 10 and Na the next biggest atomic number 11, Ne and Na are not neighbours but lie on opposite sides of the table, in Groups 18 and 1, respectively.

**The periodic table of chemical elements. The element's number is its atomic number, that is the number of protons its nucleus contains. Some superheavy elements beyond meitnerium (Mt) have been observed but not yet formally named. Groups are labelled 1–18.**

# Ions and Valency

Almost everyone knows that the molecular formula of water is $H_2O$, and that of common salt/sodium chloride is NaCl; and since the advent of global warming, most of us know that carbon dioxide is written as $CO_2$. The chemical elements combine with each other in fixed proportions that vary with the element. Thus, 1 atom of oxygen combines with 2 atoms of hydrogen to make water, but 2 atoms of oxygen are required for each atom of carbon to make carbon dioxide. Or to take slightly less familiar examples, in caustic soda/sodium hydroxide, NaOH, oxygen combines with a sodium atom and a hydrogen atom, and in sulphuric acid, $H_2SO_4$, 4 oxygen atoms combine with 1 sulphur atom and 2 hydrogen atoms. Of course, the majority of molecules in living systems, such as DNA (deoxyribonucleic acid), are vastly more complex than these simple molecules. But the same principles of chemical combination apply to them.

Chemical bonding involves the sharing of electrons between atoms. At one extreme, an electron (or electrons) is almost completely removed from an atom and taken over by another atom – an ionic bond; at the other extreme, the electron is shared equally by the bonded atoms – a covalent bond. When an electron is removed during ionization, an atom becomes a positively charged ion, a *cation* (from the Greek for 'down'), and the atom that acquires the electron becomes a negatively charged *anion* (from the Greek for 'up'). NaCl is an ionic compound, which can therefore be written as a sodium cation combined with a chlorine anion, $Na^+Cl^-$. $H_2O$, on the other hand, is essentially covalent; 2 electrons, 1 from each hydrogen atom, are shared with the oxygen atom, rather than being attached to it. (In fact, water is very weakly ionized – see p. 91.)

Ionization leads to the concept of *valency*, 'the bonding propensity of each element that depends on how many electrons its atoms have "to spare"', notes Philip Ball in *The Elements*. Valency expresses the fixed proportions in chemical combination. In NaCl, Na and Cl each have a valency of 1; in $H_2O$, H has a valency of 1, oxygen a valency of 2; and in $CO_2$, C has a valency of 4, oxygen a valency of 2. Some elements can take several valencies, depending on the atoms they combine with: in carbon monoxide, CO, C has a valency of 2, while iron (Fe) displays valencies 2 and 3 in different compounds, and phosphorus (P) valencies 3 and 5. Variable valency arises because some elements are more powerful removers of 'spare' electrons than others.

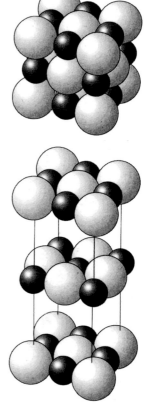

Above: **The structure of common salt, sodium chloride. The diagram at the top shows the actual structure consisting of sodium cations (small spheres) and chloride anions (large spheres). Each cation is in contact with six anions, and vice versa, as is clear from the exploded view of the structure. The photograph** (top left) **shows the characteristic cubic shape of the salt crystal.**

# Chemical Concentration

'The poison is the dose', said Paracelsus, the 16th-century physician and alchemist – meaning that toxicity depends on the concentration of a chemical. 'Even water is poisonous in excess. Even the deadly nerve gases are harmless at the level of a picogram [$10^{-12}$ g]' – this comes from the environmentalist and chemist James Lovelock, who invented the electron capture detector (ECD) in 1957, which first detected the global spread of pollutants like the pesticide DDT and the chlorofluorocarbons (CFCs), and who went on to develop his influential theory of Gaia, which is bolstered by scientific evidence collected by many scientists using the ECD.

In the news media, concentration is most commonly cited in parts per million (ppm) or parts per billion (ppb); for example, atmospheric carbon dioxide concentration stood at 381 ppm in 2005 (up from 275 ppm before the Industrial Revolution). With alcohol content, concentration is given in percentage alcohol by volume; wine is generally 11–15 per cent alcohol by volume, spirits 40 per cent or more. In the case of pollen concentration, the publicly announced pollen count refers to the number of pollen grains per cubic metre of air. Another example of concentration is the octane number of fuel, also known as its anti-knock rating. ('Knocking' or 'pinking' describes the explosive malfunctioning of an engine due to premature ignition of the fuel.) A fuel is rated by comparing its performance in a single-cylinder, four-stroke, spark-ignition engine with a blend of two reference fuels: the hydrocarbons iso-octane, which resists knocking, and heptane, which knocks readily. When the fuel under test and the reference blend are matched in performance, then the fuel's octane number is the percentage by volume of iso-octane in the reference iso-octane-heptane mixture. The higher the octane number, the less a fuel will knock (and the more expensive it is).

Scientists generally measure concentration in other units, based on the amount of substance dispersed in a solvent, which may be solid or gaseous, as well as liquid. The most common units are: the amount of substance per unit volume of solvent, and the amount of substance per unit mass of solvent. Since the amount of substance is measured (in the SI system) in moles, the scientific units of concentration are generally either moles per litre or moles per kilogram. Amount of substance depends on the number of elementary units – atoms, molecules, ions, electrons, etc. – present. *The number of elementary units in a mole is the same for all substances.* By arbitrary agreement, the mole is set equal to the amount of substance that contains as many elementary units as there are atoms in 12 grams of carbon-12, the commonest isotope of carbon. This number of units in a mole is known as Avogadro's number, after the physicist Amedeo Avogadro, who first appreciated the concept later called the mole (and himself introduced the word molecule). Avogadro's number, experimentally determined, is extremely large, just over $6 \times 10^{23}$.

Above: **Electron capture detector, invented in 1957 by James Lovelock. Its capacity to measure very low concentrations of chemicals was recognized in the 1990s by the award of major environmental prizes. Lovelock himself described it as 'the most sensitive, easily portable and inexpensive analytical device in existence'. He gives an example of its exquisite sensitivity: 'Imagine that you had a wine bottle full of a rare perfluorocarbon liquid somewhere in Japan and that you poured this liquid onto a blanket and left it to dry in the air by itself. With a little effort, we could detect the vapour that had evaporated into the air from that blanket here in Devon [Lovelock's home in the UK] a few weeks later. Within two years, it would be detectable by the ECD anywhere in the world.'**

# Acidity and Alkalinity

The acidity or alkalinity of the moisture in soil is probably its most important single property for agriculture, since it determines which plants and bacteria will grow in the soil. Food crops, in particular, prefer neutral or slightly acidic soil.

A curious fact about the house hydrangea (*Hydrangea macrophylla*) is that its blooms change colour with acidity and alkalinity. This is reminiscent of the well-known litmus test for acidity and alkalinity – except that with litmus, the colours are reversed. Litmus is a water-soluble dye extracted from certain lichens. Litmus paper, impregnated with the litmus dye, is red under acidic conditions, purple under neutral conditions, and blue under alkaline conditions. The colour change occurs over the pH range 4.5-8.3 (at 25 °C). So-called 'universal indicator' solution, containing synthetic chemicals, covers a wider pH range than litmus and the entire range of the visible spectrum: from red (very acidic) and orange/yellow (acidic), through green (neutral), to blue (alkali) and purple (very alkali).

The concept of pH was introduced in 1909 by Søren Sørensen. There is some doubt, however, about the origin of the term. It has been attributed to the Latin *pondus hydrogenii*, to the French *pouvoir hydrogène*, and in English, to 'hydrogen power', 'power of hydrogen' or 'potential of hydrogen'. At any rate, pH is an indicator of the concentration of hydrogen ions ($H^+$) in solution. An excess of hydrogen ions produces acidity. Even distilled water has a pH value, 7.0, because it contains a very low concentration of hydrogen ions ($10^{-7}$ moles per litre) produced by the very slight dissociation of $H_2O$ into the ions $H^+$ and $OH^-$. It is considered neutral, neither acidic nor alkaline – unlike many samples of drinking water, which contain dissolved minerals. The higher the concentration of hydrogen ions, the lower the pH, because pH is equal to *minus* the logarithm of the hydrogen ion concentration ($-\log 10^{-7} = 7.0$, for distilled water). On a logarithmic scale, a shift from pH 3 to pH 2 indicates a 10-fold increase in hydrogen ion concentration. (Strictly speaking, pH depends on hydrogen ion 'activity', a function of many variables, including concentration.)

| Substance | pH |
|---|---|
| Acid mine run-off | –3.6-1.0 |
| Battery acid | –0.5 |
| Gastric acid | 1.5–2.0 |
| Cola | 2.5 |
| Vinegar | 2.4–3.4 |
| Orange or apple juice | 3-4 |
| Beer | 4.5 |
| Acid rain | <5.0 |
| Coffee | 5.0 |
| Tea | 5.5 |
| Healthy skin | 5.5 |
| Normal rain | 5.6 |
| Milk | 6.5 |
| Drinking water | 6.5–8 |
| Distilled water | 7.0 |
| Healthy human saliva | 7.4 |
| Blood | 7.4 |
| Sea water | 7.4–8.2 |
| Hand soap | 9–10 |
| Bleach | 12.5 |
| Household lye | 13.5 |

Above: **A pH-sensitive flower. The house hydrangea changes colour, depending on whether its soil is acidic or alkaline. It is blue in acidic soil (pH <6.0), pink in neutral or alkaline soil (pH >6.8).**

Left: **The pH values of some common substances. Note that sea water is naturally alkaline, though its pH is 0.1 of a unit less than its pre-industrial level, as a result of increased dissolved carbon dioxide in the ocean, which forms a weak acid. Further acidification of the ocean threatens to kill organisms such as coral reefs.**

# Volume and Pressure

Even though we cannot see it, a gas exerts pressure. Familiar examples are balloons and bicycle tyres. If we squeeze an inflated balloon, thereby reducing its volume, we feel the pressure of the invisible air resisting compression. Another well-known example, which we cannot feel but can easily see evidence of, is atmospheric pressure, as measured by barometers (see p. 73). At sea level the atmosphere presses down with an average pressure of about 101,325 pascals (newtons per square metre).

Liquids, too, exert pressure, but are far less compressible than gases. Brownian movement (see pp. 83–4) demonstrates this, as does the water pressure on divers and submarines. When an object is taken underwater, the pressure exerted on it rises rapidly, at a rate of about 10,000 Pa for each metre of descent beneath the surface. At a depth of 10 metres, the pressure has doubled, compared to the surface, to 2 atmospheres. At 1400 metres down, the pressure is about 140 atmospheres, or $1.4 \times 10^8$ Pa – approximately 1 ton per square inch.

Robert Boyle, the 17th-century scientist who first investigated the relationship of pressure to volume in a gas, compared the lower strata of the air to a number of sponges or small springs; these were compressed by the weight of the higher layers of air. Boyle published his findings in a book, *The Spring of Air*, in 1660, which introduced the word elastic in its current meaning. Two years later, following experiments with an air pump designed by his assistant Robert Hooke, Boyle published a revised edition, which contained the law that is taught in school physics and is Boyle's most familiar contribution to science. Boyle's law states that the volume and pressure of a fixed amount of gas are inversely proportional at constant temperature. Halve the volume of a balloon by squeezing it, and you will roughly double its air pressure. (If, by contrast, you halved the balloon's volume by releasing some air, then of course the pressure would fall.)

For real gases, Boyle's law is followed exactly only at very low pressures, when the concentration of molecules is sufficiently low that the molecules themselves occupy a negligible volume compared to the volume of the gas. At higher pressures – like people at a crowded party in a small room – the molecules are compelled to deviate from their normal behaviour. An *ideal* gas, an important concept in science, is defined as a gas that follows Boyle's law exactly and that has, in addition, an internal energy independent of the volume it occupies. In the kinetic theory of gas molecules, this definition means that an ideal gas has negligible intermolecular attractions.

Robert Boyle (1627–91), one of the founders of the Royal Society, celebrated for his discovery of Boyle's law relating volume and pressure in gases. Boyle made many other discoveries in physics but was primarily attracted to chemistry, in which 'he was the main agent in changing the outlook from an alchemical to a chemical one' (*The Hutchinson Dictionary of Scientific Biography*). The plaque (left) to commemorate the work of Boyle and his assistant Robert Hooke appears on the wall of University College, Oxford. Thomas Young, the 19th-century polymath, wrote of this pair in 1807: 'A Boyle and a Hooke, who would otherwise have been deservedly the boast of their century, served but as obscure forerunners of Newton's glories.'

In a house on this site between 1655 and 1668 lived ROBERT BOYLE Here he discovered BOYLE'S LAW and made experiments with an AIR PUMP designed by his assistant ROBERT HOOKE Inventor, Scientist and Architect who made a MICROSCOPE and thereby first identified the LIVING CELL

# Temperature and Energy

Temperature began to be measured on the Fahrenheit and Celsius scales in the 18th century, but without understanding its scientific relationship to heat. Then, an efficient steam engine was invented by James Watt and theorized by Sadi Carnot; and Kelvin proposed the idea of absolute zero (see p. 72). In the second half of the 19th century, the kinetic theory of heat and temperature and the laws of thermodynamics were formulated. They account for familiar observations such as that a cup of hot tea spontaneously cools while a cup of cold tea never warms; that a common salt crystal spontaneously dissolves in water but never spontaneously 'un-dissolves'; and that no engine can convert fuel energy into work, its mechanical equivalent, with 100 per cent efficiency.

Basically, the 1st law states that heat and work are equivalent, and that the principle of the conservation of energy is correct for a closed system. The 2nd law – the best known – states that heat does not spontaneously pass from a cooler to a hotter body; or alternatively, that the energy in a closed system will inevitably tend to become distributed in the most probable, i.e. disordered, pattern. The 3rd law states that it is impossible to reduce the temperature of a system to absolute zero in a finite number of operations. Thus: '*1st law*: Heat can be converted into work. *2nd law*: But completely only at absolute zero. *3rd law*: And absolute zero is unattainable!' – Peter Atkins's 'sardonic summary' in *The 2nd Law*.

Temperature and energy are elusive concepts. A single atom or molecule cannot have a temperature, only an assemblage can. Even then, the temperature cannot be measured directly – it must be measured indirectly, by the height of a mercury column, for instance. When we say mercury in a thermometer is 'at' the temperature of the surrounding air, we really mean that the average energy of the mercury atoms must be neither greater nor less than the average kinetic energy of all the air molecules.

Energy, measured in joules, is not a substance but a mathematical abstraction; it makes no physical sense to speak of 'pure energy'. Its forms include kinetic energy, electrical and gravitational. Energy was first defined in this sense by Thomas Young in 1807: 'Since the height, to which a body will rise perpendicularly, is as the square of its velocity, it will preserve a tendency to rise to a height which is as the square of its velocity. The same idea is somewhat more concisely expressed by the term energy'. Hence the fact that the stopping distance of a car is proportional not to its speed but to the square of its speed, in other words to its kinetic energy.

James Watt (1736–1819), inventor of the first practical and efficient steam engine. Watt's first steam engine, built in 1769, improved on the design of Thomas Newcomen by introducing a separate steam condenser, so that the piston could be kept hot while the condenser was kept cool. He also coined the unit of horsepower for measuring engine power, by measuring the rate at which horses could work. In today's SI system, the unit of power is named the watt. 1 watt is the power when work of 1 joule is done in 1 second, or an equal heat transfer occurs in 1 second; 1 horsepower is equal to 745.7 watts.

The energy of modern civilization. This ethereal image, based on satellite photography of the Earth at night, shows the power output of city lighting in the major conurbations of the world.

# Mass and the Standard Kilogram

Movies about space, like *Apollo 13*, not to mention real-life footage of spacecraft, show astronauts and their possessions floating in zero or near-zero gravity conditions. They do not have weight, but of course their *mass* – that is their quantity of matter – has not changed at all from their mass on Earth. Free fall of bodies when spacecraft engines are switched off reminds us that while mass and weight are intimately related, they are not equivalent, although we may treat them as equivalent on Earth.

This essential point is illustrated graphically below. On Earth, 1 kilogram weighs (almost) exactly 2.2 pounds. A beam scale with 1 kg in the left pan and 2.2 lb in the right pan will balance on Earth. Indeed, the beam will balance on the Moon, on Mars and in deep space, in fact anywhere in the Universe, because gravity will act equally on the mass in each pan. But if instead we take a mass of 2 lb and place it on a spring balance, the scale readings in grams will be very different on Earth, on Mars and on the Moon – less on Mars than on Earth, and even less on the Moon (about 1/6th of the reading on Earth), because now gravity acts on the 2 lb mass alone and gravitational attraction varies considerably between Earth, Mars and the Moon (depending on the mass and radius of the heavenly body).

Of the seven basic SI units – the metre, kilogram, second, ampere, kelvin, mole and candela – the kilogram is the sole unit still defined by a man-made object, with no referent in nature. 'In a vault on the outskirts of Paris there is a small cylinder made of platinum and iridium that has the unique property that its mass can neither increase nor decrease', writes Ian Robinson, of Britain's National Physical Laboratory, in the journal *Physics World*. 'This property is not the result of new physics, however, because the object is over 100 years old. Rather, the

The Canadian standard kilogram in its storage unit. Made of platinum and iridium, it was created from the international prototype of the kilogram kept at the International Bureau of Weights and Measures in Sèvres, near Paris.

The difference between mass and weight. (See text for explanation.)

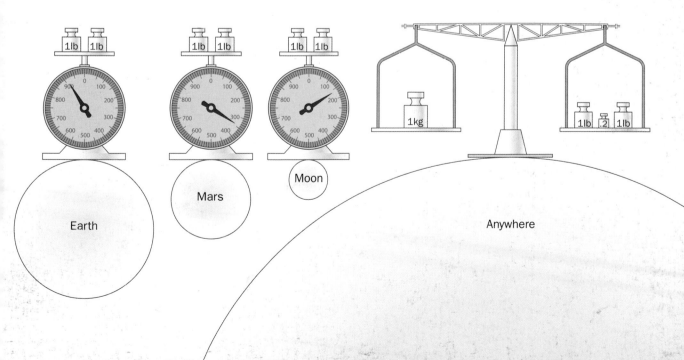

Earth

Mars

Moon

Anywhere

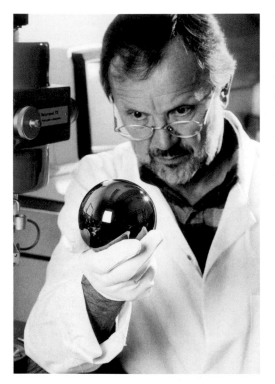

unique properties of this cylinder arise because it is the international prototype of the kilogram.' No matter how the cylinder at Sèvres has actually changed since the 1880s – through the accretion of atmospheric dirt or through diminution by cleaning – it still has a mass of precisely 1 kg, by definition. Moreover, each of the 40 copies made in 1889 for national standards institutions must have changed in mass in slightly different ways. Although they can now be compared with the Sèvres standard to an accuracy of better than 1 microgram ($10^{-9}$ kg), this is still unsatisfactory, given the unknown changes in the reference standard. Hence, the current scientific attempts to supersede the Sèvres cylinder with a universally verifiable standard.

Two basic approaches are being followed:

to count atoms and define the kilogram in terms of the fundamental constant named after Avogadro (see p. 90); or to build an electrical kilogram and define it in terms of a fundamental constant named after Planck. Both constants are known to an extremely high degree of accuracy.

With the atom counting approach, if every atom in a sample can be counted, the total number can be multiplied by the mass of the particular atom to determine the mass of the sample. In practice, this is not yet possible directly, and counting must be done indirectly via the perfection of a crystal lattice. A large sphere of silicon is used, and its diameter measured by laser interferometry (see p. 77). However, natural isotopes in the silicon have so far prevented atom counting from achieving sufficient accuracy.

The alternative, the electrical kilogram, may sound bizarre, but the idea is simple enough: design an incredibly sensitive balance that matches the weight of an object (i.e. the force exerted on it by gravity) to an electromagnetic force produced by a coil of current-carrying wire in a strong magnetic field. 'In this manner, the kilogram might be defined as the mass that can be suspended by the electromagnetic force generated when a specified amount of current flows', according to the National Physical Laboratory, which has constructed such a balance. So far, this method has achieved somewhat better accuracy than atom counting, but it still falls short of the target of measuring a kilogram with an accuracy of a millionth of 1 per cent (1 part in $10^{-8}$) on each and every occasion.

Redefining the kilogram. By counting the number of atoms in an almost perfectly spherical silicon crystal (left), or weighing a mass against an electromagnetic force (below), scientists in Australia and Europe aim to measure the mass of a kilogram to a high degree of accuracy. 'If the crystal were expanded to the size of the Earth, the difference in height between the highest peak and the deepest valley would be about 7 m' (*Physics World*). In the Watt balance at the UK's National Physical Laboratory, the cylindrical 1 kg test mass is visible just below the frame.

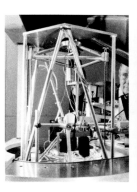

# Materials, Stress and Strain

Why are some of the chemical elements, such as oxygen, gases at ordinary room temperature, while the majority, such as sulphur (the element below oxygen in the periodic table), are solids, and mercury, uniquely among metals, is a liquid? Their forms and their physical and chemical properties can be explained by their atomic structures, at least in principle; and the same is true of all materials.

The list of properties of materials that are of practical interest is virtually endless. Density, hardness, strength, elasticity and viscosity come to mind immediately in relation to engineering. Important thermal properties include freezing and melting point, conductivity, heat capacity and expansivity. Then there are electrical properties, such as resistivity, which distinguishes metals from insulators, and semiconduction, which is essential in the silicon chip; magnetic properties that are most familiar in iron and its compounds but are widespread in nature, for example in nuclear magnetic resonance imaging; and optical properties, such as refractive index and transparency – not to forget radioactivity.

Laboratories have devised a multitude of techniques for materials testing. Perhaps the simplest is the measurement of elasticity and strength by applying a stress (a force per unit area) to produce a strain (a fractional change in length). The ratio of stress to strain is known as the modulus of elasticity. It is measured by recording the change in distance between two points on a test sample while it is being subjected to a tensile force. A strong material is able to withstand large stresses before it either shears or deforms.

**4000 BC: iron**
tensile strength, magnetic
**100 BC: concrete**
compressive strength, mouldability, durability
**50 BC: glass**
transparency, refractive properties, compressive strength
**1840s: rubber**
elasticity, water repellence, electrical resistivity
**1850s: steel**
tensile strength, hardness, processability
**1880s: aluminium**
strength : weight ratio, corrosion resistance
**1930s: polyethylene**
processability, lightness, thermal and electrical insulation, chemical resistance
**1950s: silicon**
semiconduction

Materials that have made history.

**Viscosity. Attempts to understand the viscosity of snow, and thereby to forecast the conditions likely to generate avalanches, using a water wave as a model, have not been too successful. The reason is that different types of snow have different coefficients of friction, unlike the single coefficient for flowing water.**

# Radiation and Colour

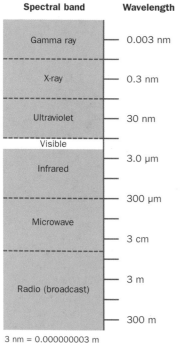

| Spectral band | Wavelength |
|---|---|
| Gamma ray | 0.003 nm |
| X-ray | 0.3 nm |
| Ultraviolet | 30 nm |
| Visible | |
| Infrared | 3.0 μm |
| | 300 μm |
| Microwave | 3 cm |
| | 3 m |
| Radio (broadcast) | 300 m |

3 nm = 0.000000003 m
3 μm = 0.000003 m

Our eyes cannot see all colours, or wavelengths, of light equally well. It is an intriguing fact – and by no means an accident – that the peak sensitivity of human vision to wavelength lies just about in the centre of the visible electromagnetic spectrum, in the yellow-green region. Moreover, this peak is quite close to the peak of the Sun's emissive power at the Earth's surface. Over millions of years, our eyes must have evolved in response to the visible spectrum and solar radiation, so that we could perceive nature as clearly as possible. It should therefore come as no surprise, as pointed out by Arthur Klein in *The World of Measurements*, that 'We humans tend unconsciously towards a kind of visual chauvinism. Radiation that we cannot "see" seems either nonexistent or an unfortunate waste.'

In reality, of course, the visible spectrum – the colours of the rainbow – is just a tiny window in the entire electromagnetic spectrum. Light is only a small portion of all radiation. Red has the longest wavelength of the colours, about 780 nanometres ($7 \times 10^{-7}$ m), yellowish-green a wavelength of 555 nm and violet the shortest wavelength, about 380 nm. Beyond red there stretches infrared radiation (heat), microwaves (used in ovens and mobile phones) and radio waves, with longer and longer wavelengths. Beyond violet stretch ultraviolet radiation, X-rays and gamma rays, with shorter and shorter wavelengths.

In terms of frequency, the longer the wavelength of electromagnetic radiation the lower its frequency, and vice versa – so infrared radiation has a frequency lower than ultraviolet radiation, and radio waves have much lower frequencies than X-rays. (The reason that frequency and wavelength are inversely related is that wavelength multiplied by frequency gives the speed of any wave; and since the speed of all

Left: **The electromagnetic spectrum. The scale is logarithmic.**

Below: **Grease spot photometer. This ingeniously simple device for measuring luminosity was invented in the 19th century by Robert Bunsen (most famous for his Bunsen burner). It relies on a screen of white paper with an oil stain. A standard candle is kept at a fixed distance from one side of the screen, and the test source (here, three standard candles) is adjusted towards and away from the other side of the screen in order to find the distance at which the grease spot becomes invisible. Too close, and the test source makes the stain bright; too far, and the stain appears dark. Luminosity is found to decline according to an inverse square law: when the distance of the test source from the screen doubles, the luminosity falls to one quarter (not one half). Hence the grease spot vanishes not when the three candles are 3 times as far from the screen as the single candle, but when they are √3 (that is, 1.732) times as far away.**

1.00

1.00

1.00

1.732

1.00

2.00

1.00

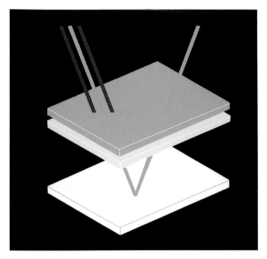

Colour perception. The wavelengths of light, and the colours of the rainbow, vary continuously across the visible spectrum, but the human eye perceives light with three discrete types of 'primary' colour receptor in the retina. First proposed in 1801 with brilliant insight by Thomas Young, this fact was not experimentally confirmed until 1959 by five American neuroscientists. In colour television, all the colours on the TV screen are created from dots of three primary colours, red, green and blue; yellow is generated by adding equal amounts of red and green light. In colour printing of magazines on white paper, green is created with an overlay of cyan and yellow, which absorb all but green light, as shown in the diagram (left). Colours must always be considered in context not in isolation; surrounding colours affect the perception of a particular colour. Colour perception is very complicated, for example the greys in this picture of grey, blue and yellow (right) appear to vary, even though they are actually identical.

electromagnetic radiation is a constant – the speed of light – if wavelength increases, then frequency must decrease, and vice versa.) According to the quantum theory, high-frequency quanta carry more energy than low-frequency quanta, so ultraviolet radiation – and still more X-rays – carries more energy per photon than infrared radiation and radio waves. This is why ultraviolet radiation and X-rays can cause cancers, because they have the energy to disrupt or break molecular bonds in human cells, whereas there is no significant health risk from exposure to lower-energy microwaves and radio waves.

The power of visible radiation is measured in watts. But since the eye is not equally responsive to all wavelengths, another unit of measurement is required to measure luminous intensity, known as the candela (from 'candle power') in the SI system. By way of comparison of the watt and the candela, the National Physical Laboratory in the UK gives the following example: 'A 60 W tungsten bulb, a normal household bulb, consumes more than 6 times the electrical power of a 9 W compact fluorescent lamp but they are both perceived as producing approximately equal amounts of light, giving out roughly the same number of candelas. This is because a lot of the power used by a tungsten bulb is given out in the infrared part of the spectrum where the eye has no response. The light given out by the fluorescent lamp corresponds more closely to the peak sensitivity of the eye.'

The candela's definition is therefore not straightforward, because it depends on the response of the eye to different wavelengths, and also because this response varies from person to person. There is no such thing as a standard visual response and we each see life in slightly different ways; for instance, the lens of the eye yellows with age. Today, after decades of experiment, a mean response from over 200 different people of both sexes covering the age range 18 to 60, is used to define the candela.

# Relativity

The world knows that Einstein discovered relativity, even if few people understand it. Besieged with requests for an explanation, Einstein eventually told his secretary: 'An hour sitting with a pretty girl on a park bench passes like a minute, but a minute sitting on a hot stove seems like an hour.'

Here we can only state Einstein's fundamental conclusions about space and time. He rejected Newton's postulate of absolute space, a tacit, universal frame of reference for the movement of all bodies, because it lacked experimental support. Instead, he postulated that the position in space of a moving body must be specified relative to a given system of coordinates (see p. 40), and that this must apply to electromagnetic radiation. But light had been measured to move at a *constant* speed (just under 300 million metres per second), whatever the movement of the observer. This made no sense without a universal frame of reference – unless absolute time too was rejected. In its place Einstein postulated that the speed of light is the same in all coordinate systems, independent of the movement of the emitter or the detector: one can never catch a beam of light, however fast one moves. Thus, the *special* relativity theory of 1905 replaced absolute, with relative, space and time.

Special relativity applies to uniformly moving bodies, not accelerating ones, so it excludes gravity. By 1916, Einstein had included gravity and created general relativity. He realized that gravity and acceleration are, in a certain sense, equivalent. It is the curvature of space – or rather *spacetime* – that is responsible for gravity, not Newton's mysterious force acting at a distance without physical contact. Matter tells space how to curve; space tells matter how to move – Einstein's grossly simplified summary of general relativity.

**Testing general relativity.** The theory's predictions have been tested many times with increasing accuracy. Indeed, the Global Positioning System relies on general relativity. The first, and most famous, test was in 1919, when a solar eclipse allowed astronomers to measure if starlight was bent by the Sun's gravity at an angle predicted by relativity. (Its success brought Einstein fame almost overnight.) In 1971, caesium-beam atomic clocks were taken aboard commercial flights and circumnavigated the globe, first eastwards, then westwards. The eastbound clocks lost time, while the westbound gained time – relative to a reference atomic clock at the US Naval Observatory. This confirmed relativity's prediction of 'kinematic time dilation': that two clocks will agree if they are at rest with respect to each other, but not if they are moving. The latest, and most accurate, experiment is Gravity Probe B (left), launched by Nasa in 2004, designed to rest relativity's prediction that the Earth will 'drag' spacetime as it rotates in space.

# Sound

The human ear is an amazingly sensitive and robust organ. At the threshold of hearing, young and healthy ears can almost detect the unceasing thermal motions in the molecules of the air; if ears were any more sensitive, we would be driven crazy by hearing the constant noise of molecular collisions. At the top end of the hearing range, near the threshold of pain, our ears can just about withstand being close to a jet aircraft taking off. The sound pressure there is one million times the sound pressure at the threshold of hearing, while its energy intensity (power per unit area) is greater by a factor of one million million, that is, the square of the sound pressure.

Sound is a mechanical disturbance from an equilibrium state that propagates through an elastic material medium. The medium can be a gas, a liquid or a solid. The disturbance is transmitted through air at a speed of about 330 m/s, through water at about 1,500 m/s, and through glass at about 5,500 m/s. But a true vacuum is silent: unlike light, sound is not transmitted through interplanetary space, because there is no elastic material medium.

A sound wave is fundamentally a longitudinal wave, in which the medium moves back and forth in the direction of propagation. In air, the molecules are successively compressed and rarefied as a sound wave passes through the air, but not permanently displaced; the medium remains in the same position after the wave moves on. The maximum temporary displacement is the amplitude of the wave, and the amplitude dictates the sound pressure.

At the threshold of hearing, the displacement is a mere millionth of a millionth of a metre (about one fifth of the radius of a hydrogen atom!) and the pressure difference between the peak and the trough of the wave is a two hundredth of a thousandth of a pascal (compare normal atmospheric pressure, which is about 100,000 Pa).

Like light, sound can be reflected, refracted, diffracted and scattered, but it cannot be polarized, because it has only one (longitudinal) component to its movement. It also has wavelength and frequency, known

**Loudness versus intensity. Intensity of a sound, measured in watts per square metre, is independent of the listener. Loudness, measured in phons and sones, depends on a particular listener's ears. Indeed, the threshold of hearing is subjective and age related. Anechoic chambers, as illustrated above, eliminate reflected sound as far as possible. They are used to compare the loudness of diffuse and progressive sound fields.**

**Doppler shift**
(see caption)

as pitch (see pp. 172–3). For humans, sound frequencies in the range 20–20,000 Hz (20 kHz) are audible, with the greatest sensitivity in the 3-octave bandwidth between 500 Hz and 4 kHz, although sensitivity to the higher frequencies declines with age and women can generally hear notes of a pitch higher than men of the same age. But animals have different frequency ranges. Dogs can hear inaudibly high-pitched whistles, and bats and dolphins use frequencies in echo-location, for hunting and communicating, that range from the humanly audible up to 150 kHz.

Frequencies above 20 kHz to about 20 MHz (though there is no theoretical limit) are known as ultrasonics, and have many applications such as depth detection (sonar), cleaning processes and the medical imaging of soft tissues by echo-location, notably of unborn babies during pregnancy (see pp. 181–2). Frequencies below 20 Hz are called infrasonics, and are in fact audible to humans if the sounds are made sufficiently intense. Long pipes capable of resonating at below 20 Hz have been used experimentally

in concerts to create audience reactions such as an extreme sense of sorrow, coldness, anxiety and even shivers down the spine. It is possible that infrasonic waves may be present in 'haunted houses' and may account for reported supernatural experiences. ('Supersonic', incidentally, refers to velocity not frequency; a supersonic aircraft travels faster than sound, with a speed measured in Mach numbers, where Mach 1 is equal to the speed of sound, Mach 2 twice the speed of sound, and so on.)

The measurement of sound intensity (power per unit area) is made on a logarithmic scale calibrated in decibels (dB). This unit, named after Alexander Graham Bell, was first introduced by the engineers of Bell Telephones in the United States in 1923 to describe not sound but electricity, in particular the attenuation of electrical power with increasing length of a telephone line (a major problem in its day in a big country like the US). The decibel scale for sound intensity has to be logarithmic because it must be able conveniently to compare sound intensities all the way from the threshold of hearing to the threshold of pain. Every increase of 10 dB (1 Bel), corresponds to a 10-fold increase in intensity. The threshold of hearing is taken to be 0 dB and the threshold of pain about 140 dB. A jet taking off lies in the region of 120 dB, that is about $10^{12}$ (1 million million) times more intense than the threshold. For comparison, the sound intensity in a library is around 35 dB.

**Bat ultrasonics. Bats emit short, high-frequency pulses of sound, which are generally inaudible to humans, and listen to the return echoes from nearby objects such as obstacles and flying insects. Direction is established by using several frequencies and detecting their relative intensities at the different ears. Target velocity is detected using the Doppler shift. This is the same phenomenon heard when a passing police car's siren rises in pitch on approaching an observer and falls as it moves away, because of the change in relative motion of source and observer** (see diagram top left). **First, the sound waves are compressed, their wavelength shortens and the apparent frequency increases, then the waves are rarefied, their wavelength increases and their apparent frequency decreases. For a constant-frequency bat, the return echo as it approaches an insect shows a positive Doppler shift, i.e. an increase in apparent frequency.**

# Electricity and Magnetism

For almost all of history, electricity and magnetism were not regarded as related. It was well known that a piece of amber, when rubbed, would become charged and pick up small pieces of straw; the word electric is derived from the Latin and Greek words for amber, *electrum* and *elektron*. Still better known – because kings and aristocrats collected them – were magnetic lodestones, which could defy gravity and pick up pieces of iron or make a compass needle point to magnetic north. But no one imagined that the electric phenomenon had much to do with the magnetic one. Even as late as 1807, the polymath Thomas Young felt able to state in his *Lectures on Natural Philosophy* that there was no immediate connection between magnetism and electricity.

By then, Alessandro Volta had invented (in 1800) the first battery, which produced a constant electrical current from a chemical reaction. Within a generation or so, electricity and magnetism would be conjoined by scientists. In 1820, Hans Christian Oersted crucially showed that an electric current in a wire could deflect a compass needle – that is, electricity could be converted into magnetism. This immediately inspired André-Marie Ampère to a series of experiments and a theory of electromagnetism that together earned him immortality, in that his name is now attached to the SI unit of electric current, the ampere (often abbreviated to amp). Ampère's most

important contributions were: his demonstration that there is a magnetic attraction between two parallel wires carrying current in the same direction but a magnetic repulsion when the current flows in opposite directions; and his idea that a magnet's power is due to the combined effect of tiny electric currents circulating in its atoms.

Finally, in 1831, Michael Faraday discovered the converse of Ampère's experiment: how to convert magnetism into electricity. Faraday wound two coils of wire around an iron bar (what we would now call a transformer), and connected the first coil to a battery and the second coil to a galvanometer, an instrument for measuring electric current. When current flowed in the first coil, no current flowed in the second. But as the current was switched off, the galvanometer gave a kick, suggesting to Faraday that current had been induced in the second coil by the collapsing magnetic lines of force in the first coil. Electromagnetic induction occurred, Faraday realized, when a magnetic field changed, not when it was constant. This was the key to the development of dynamos – where current can be generated (induced) by spinning a coil in a magnetic field – and of electric motors – where an electric coil can be made to rotate inside a magnet by alternating the direction of the current or the magnetic field. The first public electricity supply was introduced in 1881 in England for street lighting.

High-voltage power grid. High voltages are needed to ensure relatively low electric currents, and hence a manageable gauge of current-carrying wire. For safety reasons, the wires must be mounted on pylons or buried underground. Pylons generally carry warning signs (below left).

**Typical electric currents (in amperes).**

**Lightning bolt**
30,000 A
**Kettle**
10 A
**Computer**
1 A
**Household light bulb**
0.25 A
**Lethal current**
0.1-0.2 A
**Electric eel**
0.07 A
**Nerve impulses in the body**
0.000 000 001 ($10^{-9}$) A
**1 electron moving for 1 second**
0.000 000 000 000 000 000 16 ($1.6 \times 10^{-19}$) A

DANGER
ELECTRICITY OVERHEAD
MAXIMUM
SAFE CLEARANCE
4.8 METRES

Several times in this chapter, we have stated that light and other radiation such as radio waves and X-rays are *electromagnetic* waves. What does this really mean and how do we know it to be true?

In a water wave, water molecules vibrate up and down, transversely (at right angles) to the direction of propagation of the wave. In a sound wave, air molecules vibrate longitudinally (parallel) to the direction of propagation (see p. 101). Light turns out – from experiments with polarization – to be a transverse wave. It consists of varying electric and magnetic fields that together vary transversely to the direction of the light ray and also at right angles to each other, as shown in the diagram. The fields are not independent but intimately related – one cannot exist without the other – hence the need for the new concept: an electromagnetic wave. When its inventor, James Clerk Maxwell, calculated the wave's theoretical speed of propagation from mathematical equations describing the electric and magnetic fields, he was thrilled to discover that the answer coincided with the latest estimate for the speed of light from the laboratory. Maxwell thus inferred that light was probably an electromagnetic wave and published this conclusion in 1873. Not long after, in 1888, Maxwell's prediction was confirmed by Heinrich Hertz. From experiments with induction coils, Hertz showed that radio waves, light and radiant heat were all electromagnetic waves whose behaviour was described by Maxwell's equations, and that all of these waves travelled at the speed of light.

The ampere is one of the seven fundamental SI units. Other electrical units, such as the volt for electric potential and electromotive force (as in batteries), the ohm for resistance, and the farad for capacitance, are derived from the ampere. For example, the volt is defined as 'the difference of electric potential between 2 points of a conducting wire carrying a constant current of 1 ampere, when the power dissipated between these points is equal to 1 watt.' To understand this, we can think of electric current in terms of water flow from a water reservoir, and electric potential in terms of the height of the reservoir above an electric turbine. The turbine's power output will depend on both the water flow and the height of the reservoir. In purely electrical terms, power equals current multiplied by voltage. One very important practical implication of this equation is that power grids and pylons must transmit electricity across countries at ultra-high voltage – more than 600,000 V – to ensure relatively low currents. Higher currents would require thicker wires, with obvious disadvantages in cost, weight and heat loss.

Above: **Michael Faraday (1791–1867), who discovered electromagnetic induction and magnetic lines of force. His discoveries, and those of Kelvin, led James Clerk Maxwell to calculate that light must be an electromagnetic wave, thus unifying electricity and magnetism. The diagram below shows that in an electromagnetic wave the electric field and the magnetic field vibrate transversely to the wave's direction of propagation and at right angles to each other. (In a sound wave, by contrast, the vibration is longitudinal, i.e. in the direction of propagation.)**

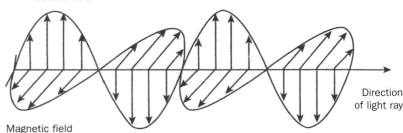

Electric field

Magnetic field

Direction of light ray

# Nanotechnology

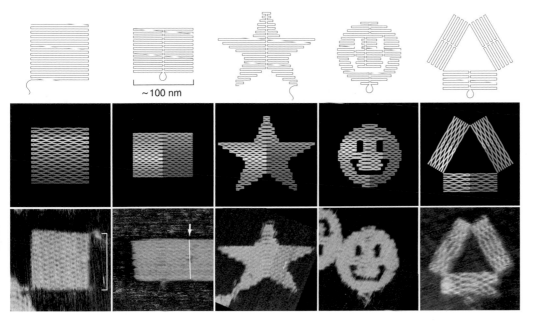

~100 nm

DNA origami shapes. (See text for explanation.)

'Why cannot we write the entire 24 volumes of the *Encyclopaedia Britannica* on the head of a pin?' asked the physicist Richard Feynman in 1959 in a talk titled 'There's plenty of room at the bottom'. His suggestion does not sound so fantastic, now that we have atomic force microscopy and scanning tunnelling microscopy (see pp. 69–70). If a water molecule were magnified to the size of a full stop, then the London Eye (the world's largest Ferris wheel) would be as big as the Earth. Nanotechnology aims to manipulate molecules and atoms as architects manipulate steel and glass. Although the high claims made for nanotechnology have yet to come true – along with dire predictions about new toxic nanoparticles – it is early days yet.

'DNA has emerged over recent years as the molecule of choice for nanodesigners', commented the journal *Nature* in 2006,

when publishing the 'DNA origami shapes' shown above, designed by Paul Rothemund. The detailed study of DNA in the half-century since its double-helix structure of interlocking base pairs was discovered (see pp. 177–8), 'allows one to predict, with reasonable success, the shapes in which a DNA molecule of a given sequence will fold in solution.'

The top row of the image shows the folding paths with the scale in nanometres. Dangling curves and loops represent unfolded sequence. The colour diagrams show the bend of helices at crossovers (where helices touch) and away from crossovers (where helices bend apart); colour indicates the base-pair index along the folding path (red is the 1st base, purple the 7000th). The bottom row shows images of real 'origami' molecules made with an atomic force microscope.

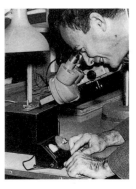

Richard Feynman (1918–88), the physicist who helped to found the field of nanotechnology. Here, in 1960, he studies what was then the world's smallest motor, 1 millionth of a horsepower, 1/6000th of an inch in diameter, built by an electrical engineer William MacLellan in answer to a challenge from Feynman in 1959.

# Chapter 6  *Earth*

Earth erupting. Satellite view on 21 July 2001 of the ash cloud erupted by Mount Etna in Sicily. The prevailing wind is blowing the ash southwards towards northern Africa, 500 kilometres from Etna. A close-up view (not shown) revealed streaks of orange near the summit of the volcano, indicating lava flows issuing from several vents. Etna has been erupting on and off for thousands of years. In 1669, one of its lava flows reached the port of Catania, 27 km from the volcano, and engulfed the streets

# Compasses

The magnetic power of lodestones has always fascinated man, especially the Chinese, who used a magnetized, swivelling, 'south-pointing spoon' in divination more than 2,000 years ago. Newton's signet ring contained a tiny chip of lodestone renowned for its ability to hold iron objects weighing 250 times its own weight. The Royal Society displayed a six-inch spherical lodestone, a terrella or 'little Earth', belonging to Christopher Wren that 'gave such Life and Merriment to a Parcel of Needles, that they danc'd the Hay by the Motion of the Stone, as if the Devil were in them', according to a normally hard-headed London journalist. In the mid-18th century, a countess, wishing to ensure the appointment of her husband as the next chancellor of Oxford University, impressed King George II with a giant lodestone grotesquely encased in a copper coronet, which she presented to Oxford's Ashmolean Museum. The lodestone apparently did the trick.

'Magnets meant money', writes Patricia Fara in her book *Fatal Attraction: Magnetic Mysteries of the Enlightenment.* Especially in Britain, which relied on seaborne trade and a powerful navy – transported by ships with magnetic compasses as a key part of their navigational equipment – to hold together its far-flung empire.

The discovery that a lodestone attached to a wooden stick and floated in water would naturally align itself in the direction of the polestar (in reality in the direction of the Earth's magnetic poles), dates from the 11th century AD in China and, possibly independently, the 12th century in Europe. The idea of replacing the lodestone with a magnetized iron needle was presumably almost simultaneous. At any rate, the first floating magnetic compass used at sea was reported from China in 1115. Subsequent improvements included the mounting of the needle on a pin standing on the bottom of the compass bowl (in the 12th century), and the mounting of the bowl on a set of gimbal rings and pivots to keep the needle horizontal in a rolling ship (around 1300). In the 17th century, the needle was converted into a parallelogram, for ease of mounting on the pivot pin. Then, in the mid-18th century, Gowin Knight invented a method of magnetizing steel 'permanently' and constructed the azimuth compass, with a bar-shaped needle and a cap to hold the needle securely on the pivot.

The major drawback of compasses in direction finding remained, though. Magnetic north differed from true north, moreover their angle of difference varied with a navigator's position because of local anomalies in the Earth's magnetic field. Furthermore, magnetic north wandered over the decades as the field changed. While navigators and scientists were aware of these inconvenient facts from as early as the 15th century, accurate adjustment of compasses to take account of them was not possible until geomagnetism was fully monitored in the 20th century.

An early Chinese mariner's compass. Chinese civilization was the first to investigate the powers and uses of magnetic lodestones.

# Surveying

Whereas location at sea was at first calculated from the heavens alone, later on from the heavens, clocks and compasses, and today depends on the satellites of the Global Positioning System (GPS), location on land came from triangulation, too. Triangulation was the basis of the first national survey in the 17th century, of the determination of the figure of the Earth in the 18th century and of the measurement of the French meridional arc that defined the metre (see p. 27).

To go into a little more detail, triangulation requires two observations of a distant station taken from two other stations through a telescope mounted on a swivelling trigonometrical instrument (a theodolite). These give two angles from the line connecting those two stations, thus making a triangle with the three stations at its vertices. Then, provided that the distance between the two stations is known, the lengths of the remaining two sides of the triangle can be calculated by trigonometry – and hence the position of the unknown third station.

Once that is fixed, triangulation can be repeated over and over, creating a network of triangles – as demonstrated below in an important triangulation between Britain and France in 1787–90, which launched Britain's Ordnance Survey, in 1791. The principal triangles were observed at night by limelight between stations at Dover Castle, Fairlight Head, Cap Blancnez and Montlambert; they had sides as long as 72 km.

Of course, stations have a height as well as a latitude and longitude, as displayed in the contours and altitude figures for mountain peaks on maps. Height measurement raises a tricky problem. What should be the zero reference point for height, the equivalent of the Greenwich meridian for longitude? Should it be at traditional mean sea level, or at global mean sea level as represented by the geoid (see p. 26), or at a level determined by the GPS, known as an ellipsoid?

Until the coming of satellites, tide gauges measured mean sea level. In Britain, the mean sea level at the fishing port of Newlyn

Below: **The beginning of the Ordnance Survey. This survey across the English Channel by William Roy in 1787–90 established the precise distance between the Greenwich and Paris observatories. Roy's baseline on Hounslow Heath, southwest of London, is highlighted, as are the official meridians of the observatories.**

Opposite top: **The survey of a subcontinent. 'Index chart to the Great Trigonometrical Survey of India, showing Colonel Lambton's network of triangulation in southern India, the meridional and longitudinal chains of principal triangles, the baselines measured with the Colby apparatus, the lines of the spirit levelling operations, the astronomical, pendulum and tidal stations, and the secondary triangulation to fix the peaks of the Himalayan & the Soolimani ranges. Completed 1 May 1870.' Note no reference to the late Sir George Everest!**

Opposite: **The Great Theodolite of the Indian survey, constructed in London around 1801, now at the headquarters of the Survey of India in Dehra Dun. The portraits on the wall show Lambton** (left) **and Everest** (right).

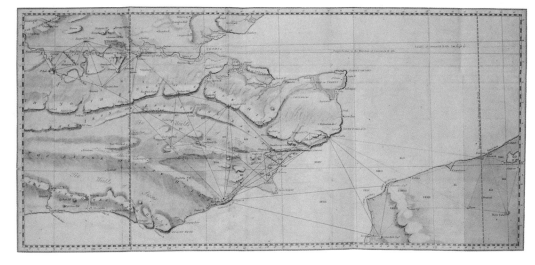

in Cornwall was determined with a tide gauge over a period of six years from 1915. Then, with the aid of this fixed point, Ordnance Survey staff worked their way across the country establishing the heights of almost 200 so-called fundamental benchmarks by the process of 'levelling'. (This consists of a long succession of short-distance sightings of the markings on a graduated measuring rod like an oversize ruler, each one of which is observed on the level, according to a spirit level, but at an elevation somewhat higher than the previous sighting. The same basic technique is employed by surveyors in construction sites.) The benchmarks were granite pillars with underground chambers set in stable ground. 'These provided a nationwide network for anyone who subsequently wished to record and plan using a three-dimensional coordinate system', notes the geodetic surveyor Marek Ziebart. The benchmarks are still used by construction engineers and also by airport authorities.

Probably the greatest land survey ever undertaken, which concluded by measuring the world's highest mountains in the 1850–60s, was the Great Trigonometrical Survey of India. Beginning in 1802 with a baseline on a plain near Madras measured by William Lambton, it concluded around 1870 in the Himalayas, having crawled 3,000 km (nearly 1,900 miles) from the tip of the peninsula to Kashmir in the far north, following the Great Arc of a meridian, with multifarious side surveys to east and west of this arc. It is a story of the indefatigable in

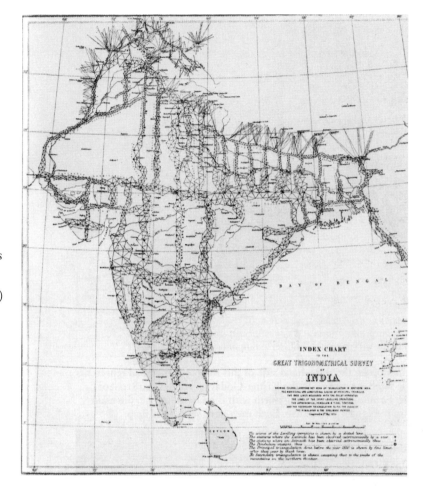

pursuit of the unmeasurable, set in a land of notorious imprecision – epitomized by the survey's driving figure, the restless, irascible surveyor-general George Everest. Curiously, his personality was the exact opposite of what the phrase 'ever at rest' suggests. Moreover Everest firmly insisted he be called 'Eve-rest', *not* 'Ever-rest'. But while Mount Everest does bear his name, although he did not survey it, most people think the name comes from its apparent meaning, not from a person – and no one at all uses Sir George's preferred pronunciation.

# Satellites

Sputnik 1, the first artificial satellite, which kick-started the space race when the Soviet Union launched it in October 1957, orbited the Earth at an altitude of about 250 km (150 miles), emitting an eerie radio beep signal that maddened American listeners. Encoded in the duration of the beep was the satellite's temperature and pressure. The only other data it transmitted to Earth concerned the density of the upper layers of the atmosphere and the propagation of radio signals in the ionosphere.

Half a century later, the 24 orbiting satellites of the Global Positioning System (GPS) of the US Department of Defense routinely pinpoint the position of a GPS receiver on Earth with an accuracy of a centimetre or less. Meanwhile Envisat, one of the European Space Agency's satellites, creates 3D maps with a topography accurate to a few millimetres from an orbital altitude of 800 km (500 miles). Such accurate remote sensing enables volcanologists to detect the deformation of a volcano before it erupts,

climate scientists to monitor the movement of ice sheets and glaciers, and town planners to model the way water runs off land so as to predict areas at risk of flooding. At the same time communications satellites – first envisioned by the science-fiction writer Arthur C. Clarke in a technical paper published in *Wireless World* in 1945 – occupy geostationary orbits, synchronized with the rotation of the Earth, in which they hover over pre-selected geographical locations. This enables them to beam telephone and television transmissions to their 'footprint' on Earth, so that we now think nothing of speaking to a person on the other side of the planet as easily as if he or she were in the same city.

The GPS and its Russian equivalent were designed in the 1970s for spying, and only later became available for scientific and

A GPS satellite in orbit. How does the GPS find locations? The location of a GPS receiver on Earth is fixed by three intersecting spheres centred on three different satellites – a method known as trilateration. When the distance between the receiver and the first satellite has been calculated, the receiver is known to lie on the surface of a sphere with the satellite at its centre of a radius equal to the calculated distance. The second satellite establishes a second sphere, which intersects the first sphere in the form of a circle. The addition of a third sphere centred on the third satellite gives two points of intersection with the circle – in other words, two possible receiver locations. Finally, a fourth satellite (not shown) with its resulting sphere distinguishes between these two intersections and fixes the receiver's location uniquely, while also allowing any time-signal errors to be measured and cancelled.

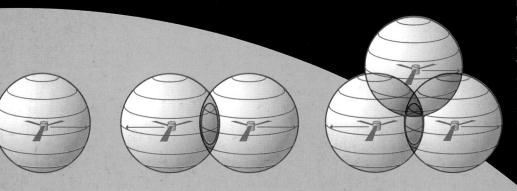

civilian use; they are still under military control with no obligation to maintain an uninterrupted service. Europe's first global navigation system, Galileo, designed in the first decade of the 21st century, is entirely independent of the military, and is intended to be reliable enough for applications in which safety is critical, such as the landing of aircraft, the navigation of cars and the running of trains. Galileo, as described by the European Space Agency, consists of '30 satellites (27 operational and 3 active spares), positioned in three circular Medium Earth Orbit planes at an altitude of 23,616 km above the Earth, and at an inclination of the orbital planes of 56 degrees with reference to the equatorial plane'. The system should provide coverage as far north or south as a latitude of 75 degrees, and be scarcely affected by the loss of any one satellite. Galileo also includes a global search-and-rescue function, capable of detecting and transferring distress signals and also of replying to the persons in distress.

Auroral activity around Earth's north magnetic pole, captured by Nasa's Polar satellite. Like iron filings bristling from the poles of a bar magnet, the magnetic field lines of the Earth's magnetic field dip down towards the planet's magnetic poles. They draw down electrons from space, which have been energized by bursts of solar activity, into the atmosphere where, at altitudes of about 400 to 100 km, the electrons excite oxygen and nitrogen molecules to states of higher energy. As these molecules fall back to their normal state, they emit photons, bursts of light, creating auroras: oxygen glows green, white or red, nitrogen blue and purple. The cause of the endlessly varied, bewitching shapes of auroras is something of a mystery.

# Weather and Atmosphere

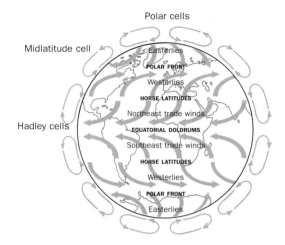

in 1735 the existence of convection 'cells', in which hot air rose high into the atmosphere at the equator, lost its humidity as it cooled, then circulated towards the poles, sank down again towards the Earth, flowed back to the equator from the north and the south, picking up moisture on the way, and was once more lofted above. However, because the Earth was not stationary but turning, the wind 'lagged behind' the Earth's surface – the nearer it was to the equator (where the Earth spins fastest), the greater the lag – making its motion appear to have an easterly component. (See diagram at left.)

Winds were classified according to their speed and the waves they produced at sea by Admiral Francis Beaufort in the middle third of the 19th century. His scale rates winds by 13 numbers ranging from 0 ('calm', sea surface 'like a mirror') to 12 ('hurricane', sea 'completely white with driving spray'). However, the first scientific steps to investigate the structure of the whole atmosphere date from the early 20th century.

Weather is produced by the interaction of the land, ocean and atmosphere with the Sun's radiation and the Earth's rotation, and is therefore complex to measure and hard to forecast. Aristotle understood that the Sun's heat evaporated moisture from the oceans, which rose into the atmosphere until it cooled sufficiently and condensed once more into water; this then fell on the oceans and the Earth as rain. Storms were thus long appreciated to be heat exchangers, taking heat from hotter parts of the Earth around the equator and distributing it in cooler parts towards the poles. The winds created acquired names according to their local characteristics, such as sirocco, foehn and harmattan.

Not until the mid-18th century, however, were the pattern and direction of the winds understood scientifically. How to account for the all-important 'trade winds' at mid-latitudes blowing from the northeast in the northern hemisphere (used by Columbus in sailing to the New World) and from the southeast in the southern hemisphere? George Hadley, a London lawyer, proposed

| Beaufort number | Description of wind | Wind speed (km/h) |
| --- | --- | --- |
| 0 | calm | <1 |
| 1 | light air | 1–5 |
| 2 | light breeze | 6–11 |
| 3 | gentle breeze | 12–19 |
| 4 | moderate breeze | 20–28 |
| 5 | fresh breeze | 29–38 |
| 6 | strong breeze | 39–49 |
| 7 | moderate gale | 50–61 |
| 8 | fresh gale | 62–74 |
| 9 | strong gale | 75–88 |
| 10 | whole gale | 89–102 |
| 11 | storm | 103–117 |
| 12 | hurricane | above 118 |

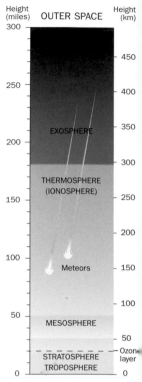

Above: **The structure of the atmosphere, which becomes thinner with increasing altitude. The ozone layer is virtually coterminous with the stratosphere, but the highest concentration of ozone is in the middle of the layer. Experiments with kites and balloons, and observations of phenomena such as meteor trails, auroras and radio-wave transmissions, led to the current division of the atmosphere into zones according to altitude.**

Above left: **Earth's winds, showing the convection cells first proposed by George Hadley, now named after him.**

Left: **The Beaufort scale.**

Clouds are almost as ubiquitous as wind in the troposphere and even in the stratosphere, except in the driest desert areas. But they are much less well understood. On the one hand there is evidence that clouds act as umbrellas and counteract global warming by reflecting solar radiation back into space; on the other, there is different evidence that they act as blankets, reinforcing the warming by trapping heat in the lower atmosphere. Moreover, our understanding of how clouds produce storms, including tornadoes, and lightning, is still nebulous. One might expect there to be more water inside a cloud than in the clear air around it, but in fact the amounts are similar. The difference is that the water vapour in the cloud has condensed into droplets, ice crystals and a mixture of ice and liquid water, because the temperature inside the cloud is lower than in the clear air. Neither is the colour of a cloud – white like cotton wool rather than black as a storm – of the expected significance. 'The colour of the cloud depends mainly on this ice/water mixture and the size of the water droplets, and less on the total amount of water', writes the physicist Albert Zijlstra.

If you want to estimate the amount of water in a cloud, the key figure is actually the volume (estimable from the size of the cloud's shadow), about a millionth of which is filled with water. A small cloud 500 m by 500 m by 100 m high – hence of volume 25 million cubic metres – contains about 25 cubic metres of water, weighing 25 tonnes.

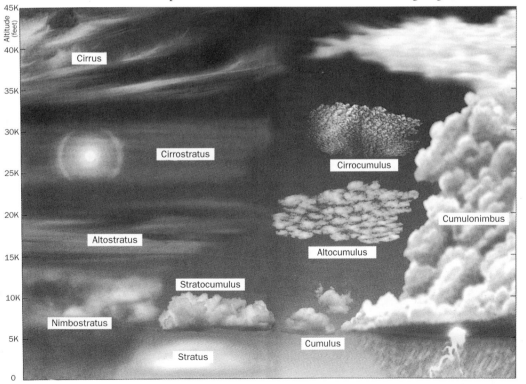

**Cloud types. The first attempt to classify clouds scientifically was made in 1802 by Luke Howard, who based it on the Linnaean system of classifying plants and animals. He coined the names cumulus, stratus, cirrus and nimbus (the last is no longer used). In 1896, there was a major revision when *The International Cloud Atlas* appeared. Cumulonimbus, that is thunderclouds, the tallest of the cloud types, were placed at number 9, and this is the origin of the phrase 'to be on cloud nine', meaning to be very happy – which caught on despite the 2nd edition where cumulonimbus moved to number 10.**

# Storms, Hurricanes and Tornadoes

At any given moment, there are roughly 2,000 thunderstorms in progress over the Earth. Together, they generate a million million ($10^{12}$) watts of lightning power – more than the combined output of all the electric power generators in the United States. One storm alone can release well over 500 million litres of water. A full-blown hurricane, 1,500 kilometres or more across and with wind speeds of up to 320 km/h, contains enough power to heat the US for half a year. The New England hurricane of 1938 altered the shape of the US east coast. Tropical cyclones in Bangladesh regularly refashion the Ganges delta (and killed hundreds of thousands of people in the 20th century). Storms can even affect the fate of nations. In 1588, the Spanish Armada was smashed by storms on the shores of the British Isles; in 1281, a typhoon in Japan left the invading forces of Kublai Khan, the Mongol emperor, at the mercy of samurai warriors, who gave thanks to Kamikaze – the Divine Wind – for the deliverance of their islands. And, at a less dramatic level, in 2005, Hurricane Katrina, by devastating New Orleans, significantly damaged the presidency of George W. Bush.

A hurricane is defined as a storm with winds of 118 km/h or more, up to a maximum speed of about 320 km/h. The name, which comes from the Carib Indian *urican*, meaning 'big wind', is restricted to storms in the Atlantic Ocean. In the Pacific, the same phenomenon is known as a typhoon (from the Chinese *taifeng*); in the Indian Ocean and around Australia, it is called a tropical cyclone. There are five levels of hurricane strength defined by the Saffir-Simpson scale (see opposite). Tornado, a

Right: **Typhoon Tip, the most intense typhoon ever measured in the Pacific Ocean, seen here almost 2,000 kilometres south of Japan and 1,450 km east of the Philippines in a satellite image. The sea level pressure in its eye, as measured from an aircraft, was found to be 870 millibars on 12 October 1979 – the lowest value on record anywhere. Besides the well-defined eye, the image shows the high degree of concentricity of the typhoon, which rotates anti-clockwise (like all northern hemisphere storms), and the spread of its great system of convection.**

Below: **Hurricane tracks, 1992–2001.**

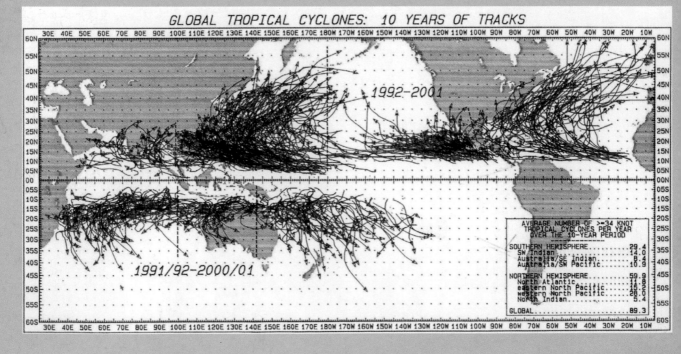

GLOBAL TROPICAL CYCLONES: 10 YEARS OF TRACKS

1992–2001

1991/92 – 2000/01

AVERAGE NUMBER OF >=34 KNOT TROPICAL CYCLONES PER YEAR OVER THE 10-YEAR PERIOD

SOUTHERN HEMISPHERE...........29.4
  SW Indian.................14.0
  Australia/SE Indian.......8.4
  Australia/SW Pacific.....10.9

NORTHERN HEMISPHERE...........59.9
  North Atlantic...........11.8
  eastern North Pacific....16.2
  western North Pacific....26.0
  North Indian..............5.4

GLOBAL........................89.3

term used worldwide, refers to a much smaller, shorter-lived and more intense storm; its wind speeds may exceed 500 km/h.

| Storm category | Maximum sustained wind speed (km/h) |
|---|---|
| Tropical storm | 50–117 |
| Hurricane: | |
|   Level 1 (Weak) | 118–153 |
|   Level 2 (Moderate) | 154–177 |
|   Level 3 (Strong) | 178–210 |
|   Level 4 (Very strong) | 211–250 |
|   Level 5 (Devastating) | 251– |

The 20th century brought scientists closer to an understanding of hurricanes (using the term to cover typhoons and tropical cyclones too) through accurate measurement, especially from aircraft and satellites – but only of certain aspects. Many of the big questions remain largely unanswered. We still have relatively little idea why some storms turn into hurricanes and others do not; why some hurricanes make landfall and others remain at sea; what causes a hurricane to wobble or even to reverse; to what extent a hurricane can steer itself; and how global

warming is influencing hurricane frequency and intensity.

From a glance at most annual hurricane track charts, there appears to be almost no pattern. But when one looks at several decades of tracks worldwide, patterns emerge. Hurricanes, typhoons and tropical cyclones originate only in specific stretches of the west Atlantic, east Pacific, south Pacific, western north Pacific, and south and north Indian Oceans. Puzzlingly, none occurs in the south Atlantic Ocean – at least none until 2004, which recorded the first ever hurricane there. They seldom move closer to the equator than latitude 4–5 °N or S, and they never cross the equator. They are much more common at certain times of year, which vary with the ocean in question. August and September is high season in the Atlantic Ocean.

The latitude condition is easily explained. Because of the Earth's rotation, air (and thus a storm) tends to rotate around depressions: the effect of the so-called Coriolis force.

Below: A cross-section of a hurricane. This vertical cross-section through Hurricane Floyd in the Atlantic Ocean in 1999 was obtained by an aircraft in the eye measuring radar reflection from rainfall. The diagram is 20 kilometres high and 120 km across; the + sign marks the location of the eye. The heaviest rain is indicated by yellow and orange. Floyd had a double eyewall structure, with inner and outer eyewalls and reduced rainfall in between – 'a fairly common feature of intense hurricanes' (Kerry Emanuel in *Divine Wind*).

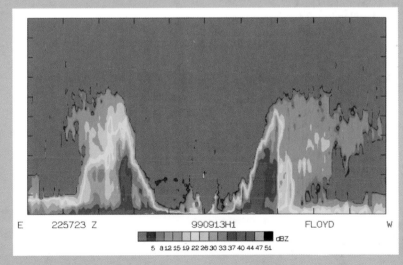

The influence of the rotating Earth on wind flow increases with distance from the equator; at the equator it is nil. In order to start spinning – and then keep spinning – hurricanes need help from the Earth's rotation; the further away from the equator they stay, the more help they get.

They also need heat from the ocean. And this fact suggests the reason for the regions in which they are born: the ocean surface must be near enough to the equator to be warm, far enough away for rotation to be induced. The threshold temperature for hurricane formation turns out to be 26 °C. If a map of the 26 °C-isotherm is drawn during different seasons of the year, hurricanes are found to originate within it. At this temperature, or warmer, winds blowing over the ocean surface can collect sufficient heat both by direct contact with the surface and ocean spray, and by the evaporation of ocean water. The heat required to vaporize the water is sucked from the ocean reservoir, which therefore cools. Satellite images of the ocean temperature before and after the passing of a hurricane off the US east coast have revealed a cooling of more than 3 °C in the open ocean, lasting more than two weeks, with even greater cooling observed in colder oceans at higher latitude or near a coastline.

Compared with tornadoes, hurricanes are quite predictable. Tornadoes can strike anywhere, especially in the midwestern US, though rarely in Africa and seldom in India. Tornadoes combine terrifyingly powerful wind speeds with vanishingly small lifetimes – often 10 minutes, maximum 2 hours – and absurdly localized tracks, as little as 50 metres wide and less than 10 kilometres in length: only 0.5 per cent of tornadoes travel more than 160 km. In shape they can resemble a long, thin rope, an elephant's trunk, or a fat, inverted bell. They think nothing of collapsing a substantial house or even carrying it away whole, picking up a school bus and dumping it upside down inside a classroom, carrying a jar of pickles 40 km unbroken, plucking the feathers off a chicken or opening the fibres of a telegraph pole and inserting a straw. They also regularly kill people – 689 in the most deadly recorded case, in three midwestern US states in March 1925.

Tornado chasing is a well-known activity there for both scientists and amateurs. But the science is still in its infancy, given the obvious difficulties and dangers of measuring tornadoes; even their maximum wind speeds are moot. Attempts have been made to drop scientific instruments into tornadoes, but most information has come from Doppler radar and photography. As photographs testify, tornado structure is remarkably complex. Very often it consists of more than one vortex; sometimes one of the mini-vortices spins clockwise, in the opposite direction to the main vortex. There are even a few instances of the main vortex spinning clockwise in the northern hemisphere, against all laws of nature. No hurricane of this kind is on record. 'We still do not have a good understanding of why storms, tornadic or otherwise, form', the meteorologist Howard Bluestein admits in *Tornado Alley*.

Opposite: **A supercell thunderstorm. Supercells are long-lasting storms that separate into updraft and downdraft regions. Beneath a rain-free part of the updraft, strong upwards motions and winds flowing inwards may produce a lowered, collar-shaped 'wall cloud'. Tornadoes frequently come out of a wall cloud, as has happened here. Doppler radar (see p. 102) can sometimes detect the birth before the tornado actually appears. 'Hook' echoes** (bottom left) **in the rainfall pattern of a storm on a radar screen are associated with tornadoes. Here, two hook echoes are visible, one at the far left and the other at the upper right. Radar can also show a vertical slice through a tornado** (bottom right), **which reveals the almost hollow 'weak-echo hole' in the centre extending from the ground to the top of the tornadic supercell 'like a miniature hurricane eye'** (Howard Bluestein).

# Lightning

A lightning flash usually lasts a few tenths of a second. It may consist of a single stroke, 3 or 4 strokes (the most common incidence), or as many as 20 or 30 strokes. The strokes are typically 40 or 50 thousandths of a second apart, hence their flicker. The current flow is generally 10,000–20,000 amperes, but may reach hundreds of thousands of amperes. It jumps in a few millionths of a second through a cloud-to-ground potential difference of several hundred million volts.

Much of the process is mysterious. We do not even know exactly why a thundercloud is tripolar, with a region of negative charge sandwiched between two positive regions. A discharge generally begins in the negative region (1) and propagates downwards as a 'stepped leader' (2–6). Near the ground it meets an upwards-moving positive charge (7) and then a 'return stroke' jumps from ground to cloud with a flash (8–10). The stepped leader is visible to special cameras, but it is the dazzling return stroke that our eyes see. It travels at almost one third of the speed of light, heats the air to some 30,000 °C and produces a 100-atmosphere pressure wave heard as thunder.

In 2002, 250 years after Benjamin Franklin flew his famous kite, scientists triggered lightning in Florida thunderstorms by launching small rockets trailing conducting wires earthed to a wooden tower. Unexpectedly, ground instruments detected bursts of X-rays. 'If humans had X-ray vision like Superman,' writes the physicist Joseph Dwyer, 'lightning would look quite different from what we are used to.' As the stepped leader forked downwards, we would see 'a rapid series of bright flashes descending from the clouds. The flashes would strengthen as they approached the ground, ending with a very intense burst at the instant the return stroke began. Although the pulse of current that would follow would be brilliant in visible light, it would look black in X-rays.' But detailed study of the X-rays has yet to produce an agreed theoretical model of lightning.

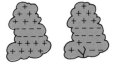

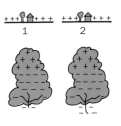

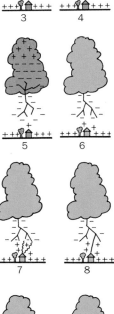

Above: **Lightning strike. (See text for explanation.)**

Left: **Rockets carrying wires into storms can trigger and measure lightning.**

# Climate Change

Ninety-nine per cent of our atmosphere – its nitrogen and oxygen – has no insulating properties. The remaining one per cent includes the so-called greenhouse gases, which act as a blanket and allow life to survive and flourish. The key gases are carbon dioxide, methane, nitrous oxide, ozone and the chlorofluorocarbons (CFCs). They absorb infrared radiation, one form of heat, in a manner similar to, though not identical with, the panes in a greenhouse.

In a real greenhouse, visible radiation enters, is absorbed by the dark plants, and is then partially reradiated by them as infrared, most of which cannot escape because the glass is opaque to longer-wavelength radiation. The atmosphere inside the greenhouse is therefore heated and its temperature rises well above that of the outside air.

In greenhouse Earth, solar radiation is absorbed by its surface and radiated back into the atmosphere as infrared, where it is absorbed by the greenhouse gases and the water vapour in clouds and then partially re-radiated. A significant proportion (twelve per cent) escapes into space (unlike in the real greenhouse), but a larger fraction radiates downwards, heating both the lower atmosphere and the surface. Measured from space by spectrometers, Earth's temperature is − 19 °C, but measured from within the atmosphere its average temperature is + 14 °C. This difference of 33 °C, which makes all the difference in the world to human beings, is the greenhouse effect.

The problem now is that the greenhouse effect is increasing and raising Earth's average temperature; and the cause is almost certainly the greenhouse gases emitted by human activities, especially carbon dioxide from the burning of fossil fuels. During the 20th century, carbon dioxide concentration rose dramatically; by the end of the century, it was about 370 parts per million, a third higher than its level of 275 ppm in 1750, at the start of the Industrial Revolution. According to scrupulously analysed temperature records, during 1900–2000 Earth's average temperature increased by 0.6 °C (with an uncertainty of 0.2 °C), and almost all the warmest years occurred in the 1990s. A major, human-made change in Earth's climate is underway. By the end of the 21st century, said the Intergovernmental Panel on Climate Change (IPCC) in 2001, the average temperature will lie in the range 1.4–5.8 °C higher than in 1990, depending on the level of carbon emissions, with potentially catastrophic effects on civilization. (See diagram on next page.)

**Charles David Keeling (1928–2005), atmospheric chemist and climatologist. In 1957–8, he began the meticulous measurements of atmospheric carbon dioxide on Mauna Loa in Hawaii and at the south pole in Antarctica that first revealed its rising concentration.**

Below: **The Antarctic ozone hole. During the 1980s, there was a steady decline in Antarctic ozone – red and yellow indicate higher concentration, blue and purple lower concentration. The cause was chlorofluorocarbons (CFCs).**

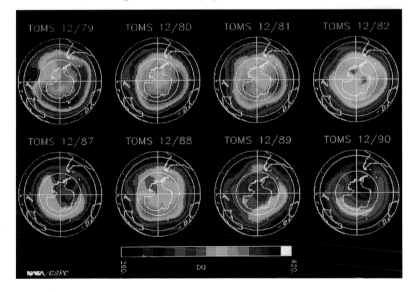

SPACE

ATMOSPHERE

Reflected by upper atmosphere 25%

Absorbed by clouds and radiated as infrared 25%

Reflected by lighter areas of Earth e.g. snowcaps

12%

Incoming solar radiation 100%

Clouds and greenhouse gases absorb infrared and radiate it into lower atmosphere

88%

**GREENHOUSE GASES**

Water vapour, carbon dioxide, methane, chlorofluorocarbons (CFCs), nitrous oxide, ozone

Outgoing infrared radiation

100%

Solar radiation absorbed by Earth's surface

45%

**HUMAN CONTRIBUTION TO GREENHOUSE GASES**

From burning of fossil fuels in power stations, factories and vehicles, from cement manufacture, from burning of tropical forests, from agriculture, from spray cans etc.

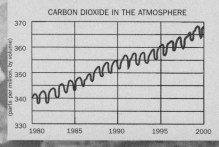

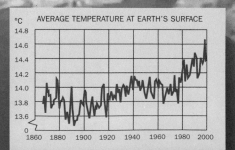

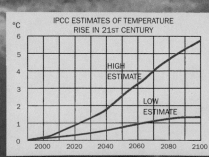

# Geological Eras

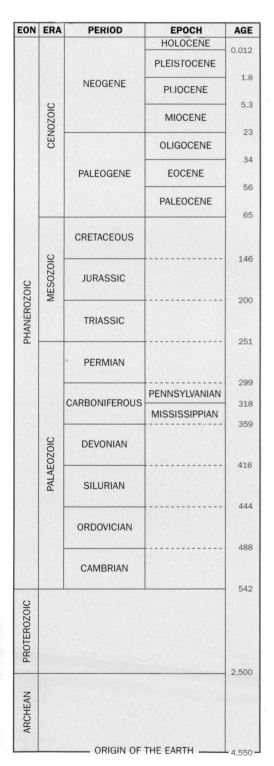

| EON | ERA | PERIOD | EPOCH | AGE |
|-----|-----|--------|-------|-----|
| PHANEROZOIC | CENOZOIC | NEOGENE | HOLOCENE | 0.012 |
| | | | PLEISTOCENE | 1.8 |
| | | | PLIOCENE | 5.3 |
| | | | MIOCENE | 23 |
| | | PALEOGENE | OLIGOCENE | 34 |
| | | | EOCENE | 56 |
| | | | PALEOCENE | 65 |
| | MESOZOIC | CRETACEOUS | | 146 |
| | | JURASSIC | | 200 |
| | | TRIASSIC | | 251 |
| | PALAEOZOIC | PERMIAN | | 299 |
| | | CARBONIFEROUS | PENNSYLVANIAN | 318 |
| | | | MISSISSIPPIAN | 359 |
| | | DEVONIAN | | 416 |
| | | SILURIAN | | 444 |
| | | ORDOVICIAN | | 488 |
| | | CAMBRIAN | | 542 |
| PROTEROZOIC | | | | 2,500 |
| ARCHEAN | | | | |
| ORIGIN OF THE EARTH | | | | 4,550 |

In the Bible, the formation of the world as described in Genesis involves seven days of creation, followed by a great flood; and the age of the Earth works out at about 6,000 years. James Hutton was the first geologist to come up with a cogent scientific alternative, in 1785, for which he is generally known as the father of geology. On the evidence of rocks in his native Scotland, Hutton proposed the principle of uniformitarianism: that the geological processes observable in the present, such as erosion and volcanism, had operated with general uniformity in the past, too. But for this to be true, immense spans of time were required, and the Earth had to be immensely old.

In the first half of the 19th century, the new science of geology divided this time span into eons, eras, periods and epochs, and created the geological column. It did so by studying strata, the order in which rocks were laid down, and fossils, the remains of living creatures in those rocks. In principle, if one stratum lay above another stratum, the top stratum must be younger – unless the two strata had been inverted by a later movement such as an intrusion of molten rock. Fossils, of course, formed recognizable groups belonging to particular time spans, which later became extinct. By the 1900s, the fundamental geological column of today was generally accepted.

Then, during the 20th century, with the advent of radiometric dating, which depends on the decay rate of radioactive elements in rocks, geologists could put absolute numbers on the geological column. For example, the

Above: **James Hutton (1726–97), father of geology, in a caricature by John Kay, 1787. Published soon after the appearance of Hutton's controversial theory of the Earth, it shows him hammering rocks that subtly resemble the faces of his critics.**

Left: **The basic geological column, as currently agreed. The duration of periods and epochs is often controversial, and is kept under regular review by geologists. (Age is measured in millions of years ago.)**

time span of the Jurassic period of the Mesozoic era was dated in 2000 as being from 146 to 200 million years ago. To be more precise, the age of the Jurassic-Triassic boundary is 199.6 million years (with an uncertainty of 0.3 million years), based on uranium-lead dating; the scientists measured the decay of two uranium isotopes into lead isotopes (see pp. 86–7) in samples of the mineral zircon taken from a layer of volcanic tuff at the top of Triassic strata on Kunga Island, off the coast of western Canada.

There are real problems with the geological column and timescale, however. First, the names given to strata in one country may be almost meaningless to geologists in another country. 'In the United States a whole suite of names has been applied to Ordovician stages that many European geologists would find as difficult to decipher as Serbo-Croat: Canadian, Chazyan, Blackriverian, Trentonian and Cincinnatian', writes a British geologist, Patrick Wyse Jackson. 'These are approximately equivalent to the Tremadoc, Arenig, Llanvim, Llandeilo, Caradoc and Ashgill stages in Britain.' Second, radiometric dates vary considerably with different rocks and different techniques – a problem which also affects the carbon dating of archaeological samples from the comparatively recent past. In 1987, the end of the Jurassic period was estimated to have occurred 131 million years ago. This was based on potassium-argon dating of the mineral glauconite, until it was later realized that argon seeps out of this mineral, making the glauconite seem younger than it really is. The currently accepted finish date of 146 million years for the Jurassic period uses potassium-argon dating of basalt – as given in the geological column on p. 121. 'The geological column', writes Jackson, 'continues to evolve.'

Dendrochronology. By counting tree rings, the age of wood can be established. This sample is from a vertical roof support in a pithouse in Broken Flute Cave, Arizona, in the southwestern United States; the researcher's pointer rests on the ring for AD 543.

Below: Section from the Cretaceous period timescale. It covers the end of the period, 65–70 million years ago, when the dinosaurs went extinct, as indicated by the fossil record.

## CRETACEOUS PERIOD TIMESCALE

| AGE (Ma) | Stage | Polarity Chron | Ammonite zones Tethyan | Ammonite zones W. Interior, N. Amer. | Belemnites and other Macrofossils | Microfossil datums Planktonic foraminifera | Microfossil datums Calcareous nanoplankton | Main Seq. T R |
|---|---|---|---|---|---|---|---|---|
| 65 | Paleogene (Danian) | C29 | | | | P1  Pα & PO | NP2 / NP1  CP1 | |
| 65.5 ± 0.3 | | | terminus | (Triceratops dinosaur fauna) Few usable ammonites | Belemnella casimirovensis | Abathom. mayaroensis, Gansserina gansseri | Micula prinsii / Nephrolithus frequens | 26 |
| | Maastrichtian | C30  U | fresvillensis | Jel. nebrascensis | B. junior | Abathom. mayaroensis | Micula murus / Lith. quadratus | CC25 |
| | | | | Hoplo. nicolleti | | | | |
| | | C31 | | Hoplo. birkelundi | B. fastigata | Racemi. fructicosa, Contusotrun. contusa | Reinhardtites levis | CC24 |
| | | | Pachydiscus neubergicus | Bac. clinolobatus | B. cibrica | | | |
| | | L | | Bac. grandis | B. sumensis | | Quadrum trifidum | |
| 70 | | | | Bac. baculus | B. obtusa | | | |

# Plate Tectonics

In 1911, while pondering the striking apparent fit between the Atlantic coastlines of Africa and South America, a versatile meteorologist and astronomer, Alfred Wegener, became convinced that continental drift had occurred. A megacontinent, which Wegener dubbed Pangaea ('all land'), had broken apart and after millions of years the bits had drifted into the present continental configuration. Unfortunately for Wegener, although his basic idea was triumphantly correct, his proposed mechanism and his calculations of the rate of movement were flawed. Moving continents were therefore rejected by a large majority of scientists until the 1960s, when the mass of diverse evidence in favour of the idea became so overwhelming that a revolution in the Earth sciences occurred.

The earliest pieces of evidence came from the floor of the Atlantic Ocean. The existence of a mountain range beneath the mid-Atlantic had been suspected since the 1850s. In 1947, scientists began to plot the shape of this Mid-Atlantic Ridge using the most powerful depth sounder then available. The ridge was found to run down the centre of the Atlantic, roughly equidistant from the coasts on either side. Dredge samples revealed that the rocks of the ridge were of volcanic origin and much younger than expected; there was also far less sediment on the ocean floor than predicted from the theory that the ocean floors had been formed early in the history of the Earth. Bruce Heezen, a geologist involved, became so fascinated that he and Marie Tharp, his drafting assistant, began to collect depth recordings from all over the world and from them to create the first profiles and three-dimensional maps of the ocean floors (see next page). When Heezen took this map and plotted on it the epicentres of Atlantic earthquakes, he suddenly realized that the earthquakes were taking place in the rift valley of the Mid-Atlantic Ridge.

The theory of plate tectonics dating from the 1960s postulates that, rather than rigid continents somehow drifting through a malleable crust, the Earth's crust consists of several large rigid plates, which are growing through volcanic action at some edges and being destroyed at others, and which are also moving across the Earth, driven by convection currents in the Earth's hot core and mantle. A plate has three basic kinds of boundary with other plates: oceanic ridges/rifts, where two plates grow; oceanic trenches, where a plate is destroyed ('subducted') beneath a second plate; and what is termed a 'transform fault', where plates are neither enlarged nor diminished. Each type of boundary produces volcanoes and earthquakes, both on land and underwater. The volcanoes of Iceland are the product of an oceanic ridge, the Caribbean volcanoes the product of a subduction zone, and the earthquakes of California's San Andreas Fault the product of a transform fault. Plate tectonics explains well the so-called 'ring of fire' around the Pacific Ocean, but it cannot satisfactorily explain volcanoes and earthquakes far from plate boundaries, such as the volcanoes of Hawaii and Italy and earthquakes in the central United States and central India.

Alfred Wegener (1880–1930), in Greenland. From 1906, Wegener conducted three scientific expeditions to the Arctic and died there. His obituaries praised him lavishly as an explorer and meteorologist, but hardly mentioned his highly controversial theory of continental drift, published in German in 1915, and subsequently in French, Swedish, Spanish, Russian and English (as *The Origin of Continents and Oceans*).

Next page: **Part of a map of the ocean floors showing the Mid-Atlantic Ridge and the East Pacific Rise, which becomes the San Andreas Fault on land. Measurement of the Mid-Atlantic Ridge shows that the Atlantic Ocean as a whole is widening at the rate of about 1.8 centimetres per year; the North Atlantic opened up about 100 million years ago on this basis.**

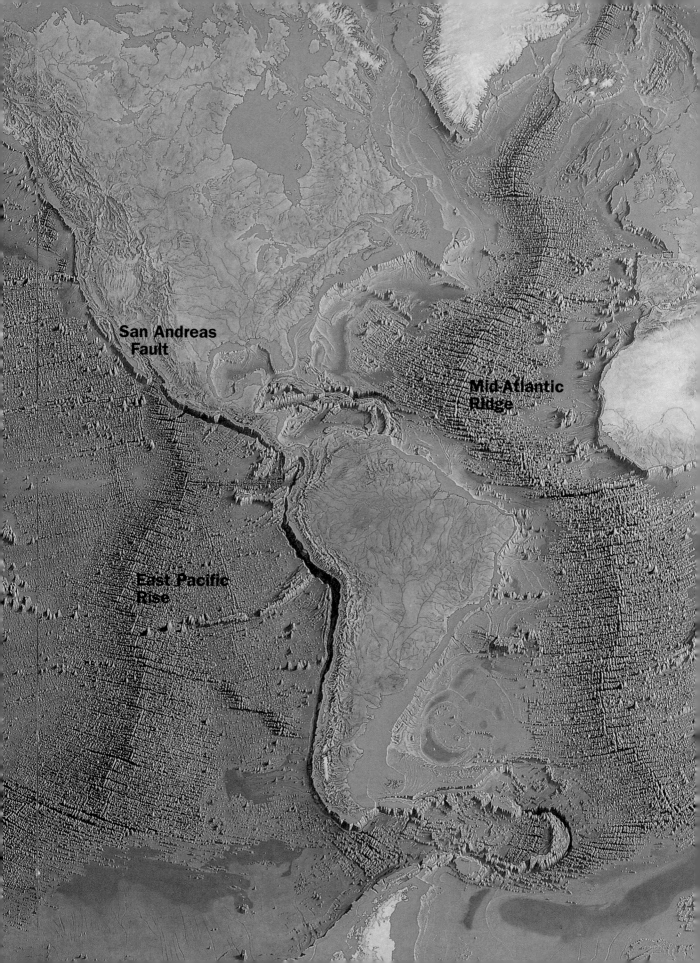

# Earthquakes

Until the mid-18th century, earthquakes belonged to the realm of the gods. The Japanese, for example, attributed an earthquake to a giant catfish, the *namazu*, said to live in the mud beneath the Earth's surface and to be restrained from pranks only by the watchful deity Kashima, who kept a mighty rock on its head. When Kashima relaxed his guard, the *namazu* thrashed impudently about.

Following disconcerting tremors in London in 1750 and the devastating Lisbon earthquake of 1755, John Michell, an astronomer and geologist, proposed that earthquakes were 'waves set up by shifting masses of rock miles below the surface': also, that there were two kinds of wave, one followed by the other; and that their speed and the centre of the earthquake could be determined by measuring the waves' arrival times at different locations. Although his second insight was not applied for almost a century, the principle is still used to determine an epicentre (surface location).

In 1857, an earthquake report from Italy reached London and attracted an engineer, Robert Mallet, to the kingdom of Naples. Assessing every crack of the damage with a trained eye, Mallet compiled isoseismal maps: that is, maps with contours of equal damage/intensity – again a method employed today, with refinements, to map seismic hazard. His maps allowed him to estimate the centre of the shaking and the relative size of the earthquake. Over twenty years, Mallet assembled a catalogue of world seismicity. It contained 6,831 listings, giving the date, location, number of shocks, probable direction and duration of the seismic waves, along with notes on related effects. The world map he created from this information, which is still accurate, was the first indication that earthquakes cluster in certain belts around the Earth.

*Intensity* of an earthquake is not to be confused with its *magnitude*, the figure generally reported in newspapers. Intensity measures the size of a quake, as does magnitude, but whereas magnitude is calculated from the vibration of a pendulum in a seismograph (see p. 74), intensity is based on visible damage to structures built by humans, on changes in the Earth's surface and on felt intensity, such as the earthquake's effect on a person driving a car. Intensity measures what human beings see after an earthquake; magnitude, what scientific instruments see.

The particular intensity scale normally used today – there are several others also in use – is a modified form of one created by the Italian seismologist Giuseppe Mercalli in 1902. It has major drawbacks. A glance at the next page will show the subjective nature of a Mercalli intensity, and that it depends on

Golden, Colorado

● Honolulu

Epicentre
(Veracruz, Mexico)

Antofagasta ●

Above: **A bullet train hit by an earthquake. In 2004, for the first time since bullet trains were introduced in Japan in 1964, one was derailed by an earthquake (magnitude 6.8) in western Japan. Nevertheless, the train was able to cut its speed from 216 to 200 km/h during a few seconds of warning received from the nationwide early-warning system for earthquakes. But in 2003, this system, with about 1,000 seismographs located around the Japanese archipelago, failed to detect any hint prior to a magnitude-8.0 earthquake off the coast of northern Japan.**

Left: **How to find an epicentre. A seismogram gives the distance of an earthquake's epicentre from the seismographic station. If three such distances are calculated from three different stations, the precise epicentre can be calculated using three intersecting circles. In this example, the epicentre of a 5.5-magnitude earthquake on 11 March 1967 lay at 19.10 °N 95.80 °W (in the sea just east of Veracruz, in Mexico).**

building construction quality, which cannot be easily assessed: one house may survive an earthquake, for instance, while an adjacent house fails. The scale is also 'culturally' dependent: damage to stone and reinforced concrete buildings is important in, say, San Francisco, but scarcely relevant in an Indian village. Finally, the scale takes no account of the observer's distance from the epicentre: a small quake close by can register a higher intensity than a large quake far away. Even so, intensity scales are extremely useful – not least for comparisons with pre-20th-century earthquakes, for which intensity is the only estimate available.

Although earthquakes have been measured from every technically feasible angle in recent decades, and despite the success of plate tectonic theory, no reliable method of prediction has developed. In 1985, the United States Geological Survey predicted a 95 per cent probability of a magnitude-6 quake at

Parkfield on the San Andreas Fault before the end of 1992. It happened in 2004! Charles Richter's words of 1958 are still true: 'One may compare [earthquake prediction] to the situation of a man who is bending a board across his knee and attempts to determine in advance just where and when the cracks will appear.'

Loma Prieta earthquake damage, Bay Bridge, San Francisco, 1989. The earthquake's magnitude was 7.1, and its intensity in San Francisco, based on the abridged Modified Mercalli scale (below, dated 1931), was XI idges destroyed').

**I** Not felt except by a very few under especially favourable circumstances.
**II** Felt only by a few persons at rest, especially on upper floors of buildings. Delicately suspended objects may swing.
**III** Felt quite noticeably indoors, especially on upper floors of buildings, but many people do not recognize it as an earthquake. Standing motor cars may rock slightly.
**IV** During the day felt indoors by many, outdoors by few. At night some awakened. Dishes, windows, doors disturbed; walls make cracking sound.
**V** Felt by nearly everyone, many awakened. Some dishes, windows, etc., broken; a few instances of cracked plaster; unstable objects overturned. Pendulum clocks may stop.

**VI** Felt by all, many frightened and run outdoors. Some heavy furniture moved; a few instances of fallen plaster or damaged chimneys. Damage slight.
**VII** Everybody runs outdoors. Damage negligible in buildings of good design and construction; slight to moderate in well-built ordinary structures; considerable in poorly built or badly designed structures.
**VIII** Damage slight in specially designed structures; considerable in ordinary substantial buildings, with partial collapse: great in poorly built structures. Panel wall thrown out of frame structures. Fall of chimneys, factory stacks, columns, monuments, walls. Heavy furniture overturned.
**IX** Damage considerable in specially

designed structures. Buildings shifted off foundation. Ground cracked conspicuously. Underground pipes broken.
**X** Some well-built wooden structures destroyed; most masonry and frame structures destroyed with foundations; ground badly cracked. Rails bent. Water splashed (slopped) over banks.
**XI** Few, if any (masonry) structures remain standing. Bridges destroyed. Broad fissures in ground. Underground pipelines completely out of service. Earth slumps and land slips in soft ground. Rails bent greatly.
**XII** Damage total. Practically all works of construction are damaged greatly or destroyed. Waves seen on ground surface. Lines of sight and level are distorted. Objects are thrown upwards into the air.

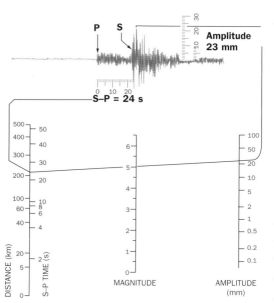

P  S  **Amplitude 23 mm**

S–P = 24 s

DISTANCE (km)
S–P TIME (s)
MAGNITUDE
AMPLITUDE (mm)

Seismologists calculate an earthquake's magnitude from the shaking recorded on a seismogram and the distance of the seismograph from the epicentre. This formula compensates for the reduction in shaking with distance of the seismograph from the epicentre. The calculations are quite tricky, and there are several scales currently in use. By far the best known is the Richter scale, published in 1935 by the Californian seismologist Charles Richter, originally to measure only local earthquake magnitudes in southern California but later modified to apply globally. (Newspapers often report an earthquake's magnitude as having been measured on the 'Richter scale', whether or not it really has been – an imprecision that maddens seismologists.)

According to the original definition, the magnitude is the logarithm of the maximum seismic-wave amplitude (in thousandths of a millimetre) recorded on a standard Wood-Anderson seismograph at a distance of 100 km from the earthquake epicentre. The reason why the scale has to be logarithmic is that the size of earthquakes varies enormously, which would make a linear scale unwiedly to use. Thus an increase in Richter magnitude of 1 corresponds to a 10-fold increase in amplitude of shaking. An earthquake of magnitude 8 shakes the ground 10 times more than one of magnitude 7, and 100 times more than one of magnitude 6. Nevertheless, a magnitude-6 earthquake may be more intense and destructive than a magnitude-8 earthquake, if its epicentre happens to coincide with a heavily populated area.

Richter magnitudes correspond to a logarithmic scale of energy released by an earthquake. For every increase in magnitude of 1, the seismic energy increases by a factor of about 30. The figure for energy released allows an earthquake to be compared with a nuclear explosion or a volcanic eruption.

Left: **How to calculate a Richter magnitude of a local earthquake. The distance from seismograph to epicentre is determined by the time interval between the S and the P waves (see p. 74), here 24 seconds, or slightly over 200 kilometres. The maximum seismic-wave amplitude is directly measurable from the seismogram, in this case 23 millimetres. A ruler placed between these two points on the left and right scales yields a magnitude of 5.0.**

Below: **Seismic energy release. This graph shows the total number of earthquakes each year in terms of their magnitude (left-hand axis) and energy release (right-hand axis). For each increase in magnitude of 1, the energy released increases by a factor of about 30.**

MAGNITUDE

ENERGY RELEASE (EQUIVALENT KILOGRAMS OF EXPLOSIVE)

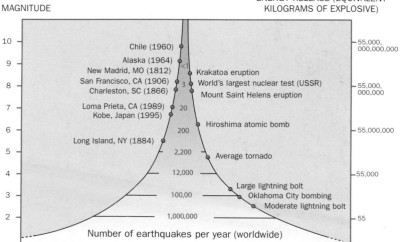

# Tsunamis

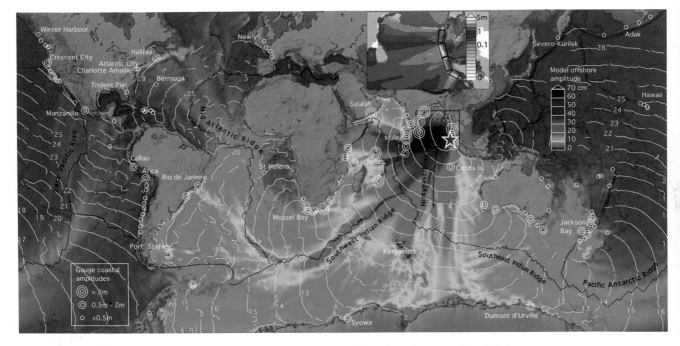

Most of the dead in the giant volcanic eruption of Krakatoa in 1883 were victims of a tsunami. It washed away 165 villages in the Sunda Strait between Java and Sumatra, killing more than 36,000. The wave reached South Africa and even as far as the English Channel.

The tsunami produced by the Sumatra-Andaman earthquake of 2004 was even more far-reaching. Almost 300,000 people died in the countries bordering the Indian Ocean, and the waves were easily detected throughout the Pacific Ocean and in the north Atlantic. For the first time, a tsunami was measured both on the surface with high-quality worldwide tide gauges and from space with multiple passes of satellites designed for altimetry. This data was used to constrain a rigorously tested computer simulation of open-ocean wave propagation known as the Most (Method of splitting tsunami) model. The colours in the global chart, created by Vasily Titov and coworkers, show the maximum computed tsunami heights during 44 hours of simulation. The contours show the computed arrival time of the waves, and the circles the locations and amplitudes of the waves in three range categories. The inset displays the four sub-faults at the source of the earthquake as constructed from satellite altimetry and seismic and geodetic data, and a close-up of wave heights in the Bay of Bengal. Most striking is the highly directional nature of the tsunami, dictated by 'the focusing configuration of the source region and the waveguide structure of mid-ocean ridges', write the chart's authors. For example, the Cocos Islands, a mere 1,700 km south of the epicentre, received less water from the tsunami than Peru and Nova Scotia!

**How the 2004 Indian Ocean tsunami spread around the world. The initial earthquake occurred at 00:59 GMT on 26 December, 100 kilometres west of Sumatra at a depth of 30 km below mean sea level. The tsunami hit Sumatra 30 minutes later, eastern Sri Lanka 2 hours later, and Africa after 7 hours and 15 minutes. (See text for fuller exploration.)**

# Volcanoes

Volcanoes come in all shapes and sizes and types of eruption. Kilauca ('Ke-low-way-ah'), a 'shield' volcano with gentle slopes in Hawaii, erupts non-explosively and produces lava fountains and relentless lava flows. These began in 1790 and have continued ever since, with the current outpouring of lava having lasted for a quarter of a century. The world-shattering eruption of the small island of Krakatoa in Indonesia in 1883, by contrast, lasted 100 days and has been followed by more than a century of comparative quiet. Krakatoa's only known previous eruption was a moderate one in 1680. Mount Etna, a substantial peak in Sicily, falls somewhere in between these two, with quite frequent periods of violent eruption producing lava flows, alternating with relative inactivity. Reports and legends about Etna's activity go back about 3,000 years. Such variety in volcanic activity, combined with an extremely patchy historical record of eruptions and the technical difficulties of getting close to an erupting volcano, make volcanoes harder to classify and more difficult to measure than earthquakes (though somewhat easier to predict).

Perhaps the most famous volcano of all, Mount Vesuvius, is a small mountain, only 1,280 metres in height, rising above the Bay of Naples in southern Italy. Vesuvius is young, scarcely 17,000 years old. It last erupted in 1944 (during the Allied invasion of Italy in the Second World War), before that in 1906, and during the past 2,000 years it has erupted more than 50 times: once every 40 years or so on average. But the actual intervals between eruptions – the repose time, in scientific parlance – bear little resemblance to the average. There was an eruption in 1037, after which the volcano slept for 600 years. When it finally awoke, in 1631, in three days it killed more than 4,000 people living in villages on its slopes, interring them in mud, ashes and lava. Naples itself, some 16 kilometres distant, was knee-deep in ash. At Portici, a town on the coast just south of Naples at the very foot of the destroyer, which had been completely ruined, the viceroy had a memorial tablet prepared and erected. It read, in part: 'Children and children's children. Hear! ....

Plinian and Ultra-Plinian

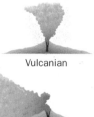

Vulcanian

Pelean

Strombolian

Hawaiian (shield)

Icelandic (fissure)

Above: **Different types of eruption and** (left) **the volcanic explosivity index. The types are named after particular volcanoes, except for the most explosive types, which are named after Pliny the Younger. The index attempts to be as scientific as possible by quantifying the eruption types by volume of tephra (volcanic fragments explosively ejected) and cloud column height above the crater and above sea level.**

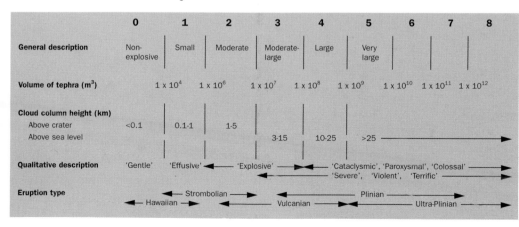

| | 0 | 1 | 2 | 3 | 4 | 5 | 6 | 7 | 8 |
|---|---|---|---|---|---|---|---|---|---|
| **General description** | Non-explosive | Small | Moderate | Moderate-large | Large | Very large | | | |
| **Volume of tephra (m³)** | | $1 \times 10^4$ | $1 \times 10^6$ | $1 \times 10^7$ | $1 \times 10^8$ | $1 \times 10^9$ | $1 \times 10^{10}$ | $1 \times 10^{11}$ | $1 \times 10^{12}$ |
| **Cloud column height (km)** | | | | | | | | | |
| Above crater | <0.1 | 0.1-1 | 1-5 | | | | | | |
| Above sea level | | | | 3-15 | 10-25 | >25 ——————————————→ | | | |
| **Qualitative description** | 'Gentle' | 'Effusive' ◄——— 'Explosive' ———► | | | 'Cataclysmic', 'Paroxysmal', 'Colossal' ———► | | | | |
| | | | | ◄——— 'Severe', 'Violent', 'Terrific' ———► | | | | | |
| **Eruption type** | | ◄——— Strombolian ——→ | | ◄——————— Plinian ———————► | | | | | |
| | ◄—— Hawaiian ——► | | ◄——— Vulcanian ———► | | ◄————— Ultra-Plinian —————► | | | | |

Sooner or later this mountain takes fire. But before this happens there are mutterings and roarings and earthquakes. Smoke and flames and lightning are spewed forth, the air trembles and rumbles and howls. Flee so long as you can … If you despise it, if goods and chattels are dearer to you than life, it will punish your recklessness and greed. Do not trouble about your hearth and home, but flee without hesitation.'

By a twist of fate, it was this eruption of 1631, now almost forgotten, that made Vesuvius a household name. Portici had to be rebuilt, and in due course, wells were sunk. Ancient Herculaneum was discovered beneath; Portici had been built on top of the ancient city's port. The discovery of Pompeii, several miles further down the coast (and also at the foot of the volcano), soon followed. Both cities had been smothered in mud, ashes and fragments of solid rock (but not lava) by an eruption of Vesuvius in AD 79,

and all but lost to memory. Were it not for a contemporary letter written by Pliny the Younger, describing his uncle Pliny the Elder's death in the eruption, the 18th-century excavators of Herculaneum and Pompeii would have known nothing about the cities.

Among those attracted to Vesuvius was Sir William Hamilton, Britain's ambassador to the court of Naples from 1764 to 1800 (and the husband of Emma, the mistress of Admiral Nelson). Vesuvius erupted violently nine times during Hamilton's stay; he made more than 200 sorties up its flanks, and became one of the pioneers of volcanology. It was he who began to compile a list of the dates of eruptions, by collecting the dates on which the priests in Naples and the villages and towns around the volcano had displayed their sacred images. Hamilton's letters to the Royal Society in London, published as *Observations on Mount Vesuvius,*

Fields of fire. Sir William Hamilton, Britain's ambassador to Naples, conducts the king and queen of the Two Sicilies and the court circle to see a lava flow on the flank of Vesuvius in May 1771; the erupting summit is seen in the background. This hand-coloured engraving by Pietro Fabris (who features himself in the foreground) is from Hamilton's famous *Campi Phlegraei* (1776), a pioneering work of volcanology.

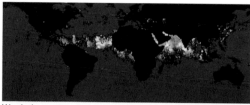
Week 1

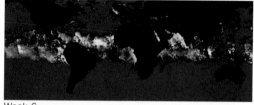
Week 6

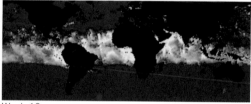
Week 10

*Mount Etna and Other Volcanos*, became the first modern work of volcanology.

More than two centuries later, despite the astonishing technology of the 21st century, eruption prediction, like earthquake prediction, remains a humbling science. Monitoring restless magma beneath a volcano is one thing; forecasting when it will erupt quite another. The single most important aspect of monitoring is continuity. The history of a volcano, both its geology and its eruptions, if well enough known, is often a good guide to its future behaviour.

Among the indicators measured by volcanologists are seismicity, tumescence and tilt, chemical emissions, gravitational, magnetic and electrical changes, and infrasonics. Of these the first, seismicity,

is the most important, because earthquakes almost invariably precede eruptions. When magma forces its way to the surface, perhaps through cracks in the crust (scientists have little exact idea), it produces tremors, often in swarms, easily measurable with seismographs. Sometimes there are major premonitory earthquakes: one such damaged Pompeii in AD 62, another damaged an important lighthouse near Krakatoa in 1880; a magnitude-7.2 earthquake in Hawaii in 1975 was quickly followed by an eruption of Kilauea that changed the volcano's pattern of behaviour. But it may be a few hours, or it may be a year or more after earthquakes begin, before an eruption gets underway.

When the magma has risen very near the surface, it bloats a volcano and deforms its surface: a fact first observed in the early 1900s in Japan and Hawaii. Eruption may follow within minutes, or it may take days. Satellites of the Global Positioning System are now key instruments in measuring the topography of volcanoes. Formerly, hundreds of reference points on the volcano's surface had to be fixed by standard land-survey techniques, and the deformation measured in relation to them. Now, GPS receivers distributed across the volcano pick up radio transmissions from satellites and read out changes in their positions (both horizontal and vertical) to an amazing accuracy, as fine as a few millimetres. At Kilauea, probably the most studied volcano in the world, GPS monitoring allows subtle changes in shape to be seen, which in turn reveal the depth and volume of magma.

**The greatest volcanic eruption of the 20th century.** The eruption of Mount Pinatubo in the Philippines in 1991, after lying dormant for 600 years, had a volcanic explosivity index of 6. Apart from huge quantities of ash, it blasted 20 million tons of sulphur dioxide into the stratosphere. The satellite images show how the sulphuric acid aerosol had spread around the world 1 week, 6 weeks and 10 weeks after the eruption. It reflected solar radiation back into space and cooled the planet significantly for two or three years, more than masking the global warming trend of the 1990s; but by 1994 average global temperature had returned to its former level. Before the eruption, volcanologists had measured the sulphur dioxide production on the ground and used its rate, along with the evidence from seismicity, to forecast that an eruption was imminent.

Chemical emissions can also be measured by satellite, but only if present in large enough quantities and after an eruption occurs – for example, the massive emission of sulphur dioxide in the 1991 eruption of Mount Pinatubo in the Philippines. To scan for escaping gases before an eruption still requires volcanologists to take samples from vents in the volcano. The chemical analysis, however, has generally to be done in the laboratory because the equipment needed is too delicate and bulky to be transported to the volcano – which means expense and

delays. So the technique has limited usefulness in prediction. Nevertheless, at Pinatubo, sulphur dioxide detection proved worthwhile. In mid-May 1991, the volcano was emitting about 500 tons of the gas daily. Two weeks later, the rate was ten times that. But then it dropped abruptly to only 280 tons daily. The monitoring team attributed this to a blockage, which pointed to a build-up of pressure underground. By then, the foci of earthquakes had closed in on the summit. The main eruption began within a week on 12 June.

**Modern instruments for measuring volcanoes on the spot include** (above) **laser geodimeters that can survey the degree of inflation of the volcano's surface, and temperature sensors made from materials such as silicon carbide able to withstand temperatures as high as 500 °C. The graph** (left) **shows Kilauea's summit tilt (tumescence and subsidence), 1956–86. As magma rises in the volcano, the magma chamber beneath the summit expands and its angle increases. After each eruption (shown by the black arrows), the angle decreases. The overall relationship between the summit swelling and eruptions is not straightforward, however, and the plumbing of Kilauea is fascinating to volcanologists.**

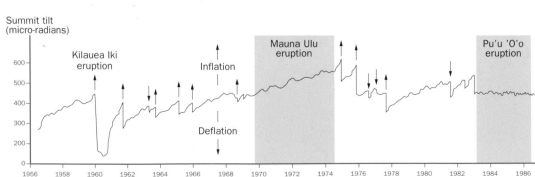

Summit tilt
(micro-radians)

Kilauea Iki
eruption

Inflation

Deflation

Mauna Ulu
eruption

Pu'u 'O'o
eruption

600
500
400
300
200
0

1956  1958  1960  1962  1964  1966  1968  1970  1972  1974  1976  1978  1980  1982  1984  1986

# Minerals, Diamonds and Gold

There is no unit for hardness of materials in the SI system, because hardness is not a clearly defined physical variable like atomic weight. Yet everyone agrees that the hardest known material is still diamond. This is why diamond – mainly the synthetic diamond available since the 1950s – is used for cutting tools and for testing the hardness of other materials.

Of the several hardness ratings used in physics and engineering, the Rockwell number measures metal hardness. A cone of diamond is pressed into the metal under standard conditions of loading; the Rockwell number is defined by the depth of the dent – the cone's degree of penetration – with a letter after the number designating the size of cone used. Another hardness scale, mainly used for distinguishing rocks and minerals in the field, is the Mohs' scale, invented in 1812 by Friedrich Mohs, a German geologist. It is a scratch test with diamond at the top, hardness 10, and talc at the bottom, hardness 1. In between come corundum (9), topaz (8), quartz (7), feldspar (6), apatite (5), fluorite (4), calcite (3) and gypsum (2). Quartz, for example, can scratch feldspar, but not vice versa; ditto calcite versus gypsum. By way of useful comparison, a steel needle is of hardness 6.5, a glass plate 5.5, a pocket knife 5.1, a copper penny 3.2 and a fingernail 2.2. Gold, depending on its alloy content, has a hardness of 2.5–3 – so a fingernail will not scratch gold.

The carat – a word derived from the Arabic *qirat*, the seed of the so-called coral tree – is the unit for measuring both the weight of a diamond or other gemstone and the purity of gold. (To avoid confusing the two meanings, the gold carat is spelt 'karat' in the United States.) One carat of a gemstone equals 0.2 g, or 200 mg, so a 5-carat diamond weighs 1 g. By contrast, pure gold is defined as being 24 carat, while 18-carat gold is 3 parts gold to 1 part alloy, 12-carat gold is half gold and half alloy, and so on.

How to tell gold from alloy was discovered by Archimedes (see p. 50). How to distinguish natural from synthetic diamonds is more problematic. The key test relies on the fact that the impurities and growth structures of the two types are different because the synthetic diamond is grown in a matter of days, as against perhaps billions of years for the natural diamond. When both are exposed to laser or filtered light, they fluoresce differently. A physicist at a diamond trading company, Simon Lawson, describes the result (see photographs): 'When viewed under intense ultraviolet light, synthetic diamonds have blocky patches of fluorescence with well-defined geometrical shapes, whereas natural diamonds have more wavy or 'tree ring' type structures.'

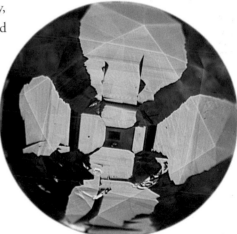

**Natural versus synthetic diamonds. An ultraviolet fluorescence testing instrument known as DiamondView can distinguish natural diamond** (top) **from synthetic diamond** (above).

# Biological Species

How to define a biological species, and how new species arise, are central questions in the life sciences – as witness the title of Charles Darwin's revolutionary work, *On the Origin of Species,* published in 1859. By then, Carl Linnaeus's binomial system of classification, introduced in the mid-18th century and still used today, was well established among both botanists and zoologists.

In the Linnaean system, the first part of the Latinized scientific name is the generic name or genus, and the second part is the specific name or species; both are italicized. For example, in the name of the common frog, *Rana temporaria*, the genus is *Rana* and the species is *temporaria*. With the creeping buttercup, *Ranunculus repens*, as opposed to the meadow buttercup, *Ranunculus acris*, there are two species, *repens* and *acris*, which belong to the same genus *Ranunculus*. The discoverer's name may be added to the binomial name in an abbreviated form in roman type, for instance, the scientific name of the common daisy is *Bellis perennis* L. (where L. stands for Linnaeus). Thus, whereas the common name of an animal or plant puts the species first (e.g. 'creeping') and the genus second ('buttercup'), the scientific name reverses this order.

The same plant may have many common names, but it has only one scientific name; in Britain, the marsh marigold (*Caltha palustris*) has 80 names. And of course one common name may be used to refer to many different plants. Consider the shamrock. The horticulturalist Anna Pavord notes amusingly in *The Naming of Names: The Search for*

*Order in the World of Plants*: 'When, in 1892, Nathaniel Colgan of Dublin tried to establish the true identity of the shamrock, patriotic Irishmen from twenty different counties inundated him with plants. Some sent white clover, some red, some sent lesser yellow trefoil, some spotted medick. No one sent wood sorrel, which in England is sometimes called shamrock.'

More general than the genus are: the *family*, the *order* and finally the *division* – all of which have been regulated since 1867 under the International Code of Botanical Nomenclature. The family Liliaceae includes lilies, tulips, fritillaries and erythroniums. The order Ranunculales groups together the buttercup family (Ranunculaceae), the barberry family (Berberidaceae) and the akerbia family (Lardizabalaceae). A division is an overarching category distinguishing, say, flowering plants from ferns or mosses.

In practice, families, orders and divisions were constantly under review, even before the coming of molecular biology in the 1950s. Now, the whole concept of species has been altered by the extraordinary specificity of gene sequencing. 'When you have seen one species of chrysomelid beetle, you have emphatically not seen them all. In fact you still know very little about the family

The earliest known illustration of a plant, from a fragment known as the Johnson Papyrus, *c.* AD 400. Labelled Symphyton, it shows *Symphytum officinale*, of the borage family, commonly known in England as comfrey.

| Approximate Numbers of Plant Taxa and Fungi | |
|---|---|
| **Taxon species** | |
| Prokaryotes (bacteria, cyanobacteria) | 3,600 |
| Eukaryotic algae | 33,000 |
| Green algae | 7,000 |
| Diatoms | 6,000 |
| Red algae | 4,000 |
| Brown algae | 1,500 |
| Fungi, including slime moulds | 90,000 |
| Ascomycota (sac fungi) | 30,000 |
| Basidiomycota | 30,000 |
| Mosses | 26,000 |
| Lichens | 20,000 |
| Fernlike plants | 15,000 |
| Lycopodiophyta | 400 |
| Horsetails | 32 |
| Spermatophytes (seed-bearing plants) | 236,000 |
| Gymnosperms | 800 |
| Angiosperms | 235,000 |

| Approximate Numbers of Species in Selected Major Animal Taxa | |
|---|---|
| **Taxon species** | |
| Protozoa | 40,000 |
| Sponges | 5,000 |
| Cnidaria (e.g. coral, jellyfish) | 10,000 |
| Flatworms | 16,100 |
| Roundworms | 23,000 |
| Molluscs | 130,000 |
| Snails | 85,000 |
| Clams | 25,000 |
| Cephalopods (e.g. octopus) | 600 |
| Segmented worms (Annelids) | 17,000 |
| Arthropods | >1,000,000 |
| Arachnids | 68,000 |
| Crustaceans | 50,000 |
| Insects | 1,000,000 |
| Dragonflies | 4,700 |
| Cockroaches | 4,000 |
| Locusts, grasshoppers | 20,000 |
| True bugs | 73,000 |
| Beetles | 350,000 |
| Bees and ants (Hymenoptera) | 110,000 |
| Butterflies | 120,000 |
| Flies, gnats | 120,000 |
| Echinoderms (e.g. sea urchin) | 6,500 |
| Vertebrates | 46,500 |
| Fish (including lampreys) | 20,600 |
| Amphibians | 3,300 |
| Reptiles | 6,300 |
| Birds | 8,600 |
| Mammals | 3,700 |

Above: **'Yeti crab' (*Kiwa hirsuta*). The discovery in 2005 of this hairy-clawed, blind crab, about 15 centimetres long, necessitated the creation of a whole new taxonomic family, Kiwaidae, distantly related to the hermit crabs, Paguridae. Kiwa is the goddess of crustaceans in Polynesian mythology. The crab was found at a depth of about 2,200 metres living on a recent lava flow in the East Pacific Rise south of Easter Island. Like many other creatures, the Yeti crab obtains nutrition from the hydrothermal vents in the ocean floor. How it does so is not yet clear, but one possibility – given the large colonies of bacteria found in the crab's hairs (which are in fact flexible, hair-like spines known as *setae*) – is that the crab 'farms' the bacteria, perhaps as a source of food.**

Chrysomelidae', writes the prominent biologist (and ant specialist) Edward O. Wilson in *In Search of Nature*. 'In its genes each species contains on the order of a million to a billion bits of information, assembled by an almost inconceivable number of events in mutation, recombination and natural selection during an average life span, according to taxonomic group, of 1 to 10 million years.'

Wilson and others estimate there are 1.4 million named species in the world, including animals, plants and micro-organisms, but that the total number of species could be anywhere between about 5 and 100 million. These figures are staggering enough, yet they may represent less than 1 per cent of all the species that have ever existed. 'We have only begun even a superficial exploration of life on Earth, living and extinct', Wilson has long maintained.

# Chapter 7    *Universe*

Armillary sphere.
It shows the great
circles of the
heavens – such as
the horizon, the
meridian, the
equator and the
polar circles – with
the Earth at the
centre. Invented by
the ancient Greeks,
the armillary sphere
takes its name from
the Latin *armilla*,
meaning 'bracelet'.

# The Heliocentric Universe

To Aristotle and Ptolemy, it was obvious that the Earth must be the centre of the Universe and that the Sun and planets must move around it in circles, since the circle was self-evidently the most perfect geometrical form, and the heavens were perfect. In fact, as we now know, the Earth and other planets orbit the Sun, and they have elliptical orbits. But, warns Alex Hebra in *Measure for Measure*, 'Rather than belittle such ancient wisdom, we moderns should recall that from later antiquity and all through the Middle Ages, the concept of circular motions led to accurate predictions of the Sun, the Moon, and the planets, and that accurate calendars were compiled and solar and lunar eclipses were predicted correctly. This is no mean feat, especially as [these astronomers] regarded the Sun as travelling around the Earth.'

One of the astronomical instruments that made this possible was the armillary sphere. A skeleton celestial globe, the sphere consisted of circles divided into degrees for angular measurement. It was immensely popular in the Renaissance as a symbol of wisdom and knowledge – either suspended, mounted on a stand or affixed to a handle. Among the armillary spheres designed to be used were some made by Tycho Brahe, in the late 16th century. Brahe developed several forms of sphere, each intended for a specific astronomical task. Although they became simpler and lighter in construction – so that they did not sag under their own weight – their intricacy reflected astronomers' attempts to account for the observed zigzags and elliptical paths of the heavenly bodies.

Armillary spheres remained popular in the 17th and 18th centuries, when they were often used to demonstrate the difference between the old, Earth-centred, Ptolemaic universe and the new, Sun-centred, Copernican universe.

The practical achievements of medieval astronomy, and the fact that the Catholic church supported Aristotle's philosophy, combined with the natural human tendency to assume that humans must be at the centre of the world, together account for why it took so long for the new vision of the solar system to emerge and be accepted. Copernicus published his heliocentric *De Revolutionibus Orbium Coelestium* on his deathbed in 1543 – but books advocating a geocentric universe continued to be published until after the end of the *17th* century. Indeed, only with the arrival of relativity at the beginning of the 20th century was the debate finally rendered meaningless. After this, the statements that 'the Sun is at rest and the Earth moves' and 'the Sun moves and the Earth is at rest' became simply different conventions concerning different coordinate systems, noted Einstein in *The Evolution of Physics*. 'Either coordinate system could be used with equal justification.'

Above: **The Antikythera mechanism, 2nd century BC. Found in a shipwreck off a Greek island in 1902, it appears to be a sophisticated eclipse calculator with gears. Its antiquity is astonishing.**

Below: **Copernicus's heliocentric cosmos, as depicted in Johannes Hevelius's *Selenographia*, 1647. Note that the orbits of the planets are shown as circles.**

# The Motion of the Planets

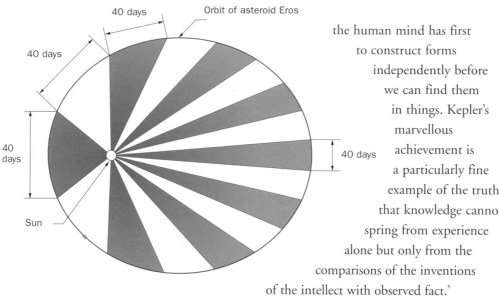

40 days

40 days

Orbit of asteroid Eros

40 days

40 days

Sun

Johannes Kepler (1571–1630), the astronomer who discovered the laws of planetary motion. His second law is shown in the diagram of the elliptical orbit of the asteroid Eros around the Sun, in which all the segments are of equal area. (See text for further explanation.)

The astronomer Johannes Kepler analysed the first accurate observational data on the planetary motions, compiled by Tycho Brahe before his death in 1601, and made an educated guess that their orbits were not circles, as believed by both Copernicus and Ptolemy, but ellipses – one of the geometrical forms 'discovered' by the Greeks. With this remarkable mental leap, Kepler conceived his three laws of planetary motion. These enabled him to calculate astronomical tables, and hence the positions of the planets at any time in the past, present or future, which fitted well with astronomers' observations, such as those of his contemporary Galileo. Later in the 17th century, Kepler's laws – especially his second law – were crucial to Newton in formulating his law of gravitation between the Earth and the Moon and between the Sun and the planets.

As Einstein remarked in 1930, on the tercentenary of Kepler's death, 'It seems that the human mind has first to construct forms independently before we can find them in things. Kepler's marvellous achievement is a particularly fine example of the truth that knowledge cannot spring from experience alone but only from the comparisons of the inventions of the intellect with observed fact.'

What we now know as the first and second laws were published in Kepler's book *Astronomia Nova* in 1609 (although the second was actually discovered first), while the third law was not published until 1620 in his *De Harmonices Mundi*. The first law states that the planets travel in elliptical orbits and that the Sun occupies a focus of the ellipses. The second law is a bit harder to follow. It states that the areas swept out by a line joining the Sun and a planet during equal time intervals are of equal size – which means that planets travel more quickly when they are nearer the Sun. The diagram makes this concept clear. The third law, the most complex, describes with mathematical precision the link between the distances of the planets from the Sun and their velocities – 'a feat that afforded [Kepler] extraordinary pleasure', notes *The Hutchinson Dictionary of Scientific Biography*, 'and confirmed his belief in the harmony of the universe.'

# The Moon

At the beginning of 2001, in his book *Patrick Moore on the Moon*, one of the world's best-known astronomers, wrote: 'Way back in 1969 it was widely believed that a fully fledged Lunar Base would be built within a few years, and that lunar tourism was looming ahead. Now, at the start of the new millennium, these aims are still a long way off.' Moore recalled talking about this in 1970 to the astronaut Neil Armstrong, who told him: 'I'm quite certain that we'll have such bases in our lifetime.' But it is now clear that Armstrong was over-optimistic, since no humans have been to the Moon for more than a generation.

This long gap has provoked comparison between lunar exploration and polar exploration on Earth. Almost half a century went by after the first voyages to the north and south poles in 1909 and 1911, driven by national rivalries, before scientists were able to work there and establish the first permanent bases during International Geophysical Year, 1957–8. If the space agency Nasa has its way, some time after 2018, half a century after the United States beat the Soviet Union to a manned Moon landing in 1969 – and Armstrong said 'That's one small step for a man, one giant leap for mankind', as he stepped onto the surface – four-person scientific teams will spend a week on the lunar surface, with a view to eventually establishing a permanent base.

'The Moon is one of the best pieces of real estate for scientific research in the solar system,' commented the magazine *New Scientist* in a special report on science on the Moon in 2006. In certain places near the poles there is perpetual sunlight, suitable for continuous solar power; and in the shadows of the crater rims, perpetual darkness, ideal for astronomy. 'The perishingly low temperatures inside craters are especially good for infrared astronomy, superconducting power systems and other types of low-temperature research.' In addition, there is a rock-steady surface on which to build structures, and the materials required to build them. 'Little wonder that scientists have been eyeing the lunar surface for so long.'

Next page: **Apollo 12 mission, 14–24 November 1969. Preparing for its final descent, the lunar module Intrepid floats 111 km (69 miles) above the giant crater Ptolemaeus, after having separated from the command module. The circular crater in the middle distance on the right is Herschel. Intrepid eventually touched down in the Ocean of Storms.**

## The Measurements of the Moon

| | |
|---|---|
| Mean distance from Earth | 384,365 kilometres (238,840 miles) or 0.0025695 astronomical units |
| Maximum distance | 406,670 km (252,700 miles) |
| Minimum distance | 356,396 km (221,460 miles) |
| Sidereal period | 27.321661 days |
| Synodic period | 29 days 12 hours 44 minutes 2.9 seconds |
| Axial inclination of equator, referred to the ecliptic | 1° 32´ |
| Orbital eccentricity | 0.0549 |
| Orbital inclination | 5° 9´ |
| Mean orbital velocity | 3,680 kph (2,287 miles per hour) |
| Apparent diameter | max. 33´ 31˝, mean 31´ 5˝, min. 29´ 22˝ |
| Magnitude of full moon at mean distance | –12.7 |
| Mean albedo | 0.07 |
| Diameter | 3,476 km (2,160 miles) |
| Mass | 1/81.3 Earth = 0.0123 Earth = $3.7 \times 10^{-8}$ Sun |
| Volume | 0.0203 Earth |
| Escape velocity | 2.38 km/s (1.5 miles per second) |
| Density | 3.34 water = 0.60 Earth |
| Surface gravity | 0.01653 Earth |

# Planets

Long before telescopes, ancient astronomers knew of five heavenly bodies, apart from the Sun and Moon, which moved appreciably against the apparently fixed background of the stars. They were the planets Mercury, Venus, Mars, Jupiter and Saturn; and their names became connected, in some languages, with the days of the week (most obviously Saturday, from the Latin *Saturni dies*, meaning 'day of Saturn'). Telescopes subsequently added three more planets: Uranus in 1781, Neptune in 1846 and Pluto in 1930 (although Pluto, as of 2006, is no longer classed as a bona-fide planet) – making nine planets.

Galileo, as we know, discovered Jupiter's four moons. He also spotted Saturn's rings, which he called 'handles'. Then, in 1659, with much better lenses and a longer telescope, Christiaan Huygens characterized Saturn's rings as a disc, and discovered Titan, its largest moon. But Huygens failed to observe Saturn's other moons – for a curious reason. He was a follower of Descartes, which encouraged his belief in numerology: that the properties of numbers revealed the structure of the heavens. A so-called perfect number is the sum of its divisors. For example, 6 is perfect: its divisors are 1, 2 and 3; 1 + 2 + 3 = 6. (The next perfect numbers are 28, 496 and 8,128.) Since the heavens were perfect, argued Huygens, they must display perfect numbering. Six planets were known, with six moons (our Moon, Jupiter's four moons and Saturn's Titan) – both perfect numbers – so there was no sense in looking for more moons.

But in 1671–2, Jean-Dominique Cassini (the founder of France's cartographic Cassini dynasty) observed Saturn's moons Iapetus and Rhea. Yet he too 'engaged in number fantasies', writes I. B. Cohen in *The Triumph of Numbers*. Keen to become director of King Louis XIV's newly constructed Paris observatory, Cassini ingratiatingly announced that the 6 planets and 8 moons of the solar system added up to an auspicious number of 14 bodies circulating around the Sun and proclaiming the glory of the 'Sun King', who was the 14th monarch of his line! Cassini got the job, although one cannot help wondering how he explained his discovery of two more moons of Saturn, Tethys and Dione, in 1684. (There are actually 40-plus moons.) Not only did he remain director until his death in 1712, Cassini is still remembered in two prominent ways: in the Cassini division (the gap in Saturn's rings that he was the pioneer in observing), and in the amazing Cassini space mission to Saturn.

Christiaan Huygens (1629–95), physicist, mathematician and astronomer, who discovered Saturn's moon Titan and the nature of Saturn's rings. The photograph below shows all that remains of his telescope: the lens, 57 mm in diameter and inscribed along the border with its focal length (10 Rhineland feet) and the date of its final polishing (3 February 1655). It also carries a verse from the Roman poet Ovid: 'Admovere oculis distantia sidera nostris.' (They brought the distant stars closer to our eyes.) Huygens is also celebrated for his work on light, in which he proposed the undulatory theory, in opposition to Newton's corpuscular theory.

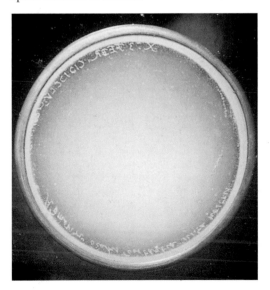

The Cassini mission has transformed our understanding of Saturn in ways that Cassini himself could not have imagined. The Cassini spacecraft, launched from Earth in 1997 on its mission (see diagram opposite), has a mass of 5,820 kilograms and is one of the largest planetary spacecraft ever built – 'the size of a school bus, even without the booms that project in varying directions', writes one of its engineers, Joan Horvath, in *Saturn: A New View*. It is also one of the most complex, not least because of the dim Saturnian sunlight (about one per cent as strong as sunlight at Earth), which precludes solar panels as a viable source of power. The photograph (right) shows the spacecraft being prepared by engineers in 'bunny suits' for testing in the thermal-vacuum chamber at the Jet Propulsion Laboratory in California, where the vibrations of launch, the intense heat of the inner solar system, and the cold of the outer solar system, can be simulated. Facing the camera is the foil-covered Huygens probe, which was released from the spacecraft on Christmas Day 2004 and landed by parachute on the surface of Titan on 14 January 2005, surviving and collecting surface data for 1 hour, 12 minutes and 13 seconds – far longer than expected. One of the mission's unique images, taken in conjunction with the Hubble space telescope, appears at right. It shows an aurora at Saturn's south pole developing over a period of five days. The space telescope took the images in ultraviolet light, while the space mission recorded radio emissions and monitored the solar wind.

Left and opposite: **The Cassini spacecraft, launched in 1997 towards Saturn.**

Below: **Saturn's rings and polar aurora, observed by the Cassini mission, in conjunction with the Hubble space telescope.**

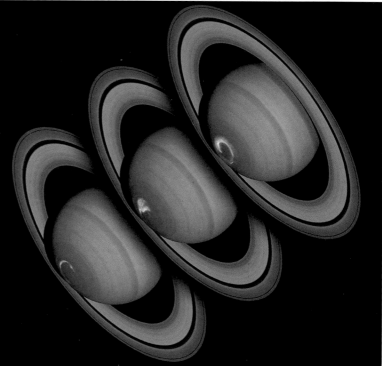

## CASSINI SPACECRAFT

Dimensions: 6.7 metres by 4 metres
Weight: 5,712 kilograms with fuel,
2,125 kilograms without fuel
Launch date: 15 October 1997

## FIELD AND PARTICLES INSTRUMENTS

Three instruments designed to map the magnetic field of Saturn and detect charged particles and plasmas. Also to study interactions between moons and the solar wind, and investigate ice, dust, plasma and radio waves

## RADAR BAY

Instruments map Titan and measure heights of surface features

## MAGNETOMETER BOOM

11-metre-long boom holds instruments to measure the size and direction of magnetic fields in Saturn's environment

## HUYGENS PROBE

Released on 25 December 2004, Huygens parachuted through Titan's atmosphere analysing its physical and chemical properties as well as taking images. Also analysed the surface

## RADIO AND PLASMA WAVE ANTENNA

## RADIOISOTOPE THERMOELECTRIC GENERATOR

Provides power for the spacecraft, including its instruments, computers and radio transmitters

## REMOTE SENSING INSTRUMENTS

Four separate instruments determine the temperature, chemical composition, structure and chemistry of Saturn, its rings, moons and their atmospheres. Also measure the mass and internal structure of Saturn and its moons. Cameras take images at visible, near-infrared and ultraviolet wavelengths

## MAIN ENGINES

Provide 445 newtons of thrust

## ORIENTATION THRUSTERS

Make small adjustments to the flight path

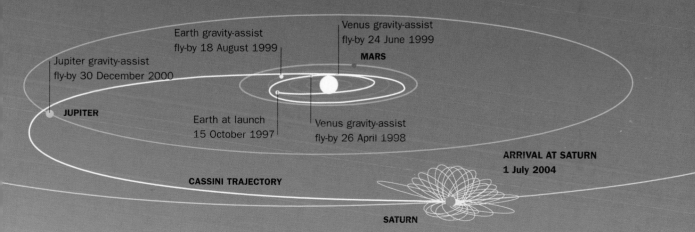

Jupiter gravity-assist
fly-by 30 December 2000

Earth gravity-assist
fly-by 18 August 1999

Venus gravity-assist
fly-by 24 June 1999

**MARS**

**JUPITER**

Earth at launch
15 October 1997

Venus gravity-assist
fly-by 26 April 1998

**ARRIVAL AT SATURN
1 July 2004**

**CASSINI TRAJECTORY**

**SATURN**

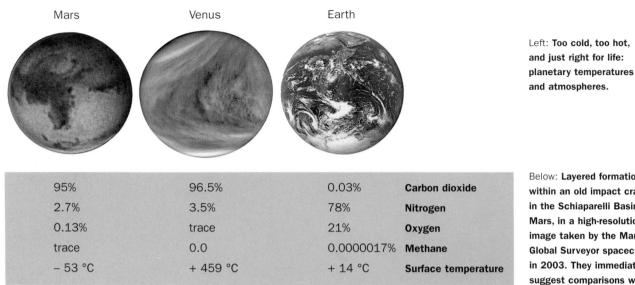

Mars　　　　Venus　　　　Earth

| Mars | Venus | Earth | |
|---|---|---|---|
| 95% | 96.5% | 0.03% | **Carbon dioxide** |
| 2.7% | 3.5% | 78% | **Nitrogen** |
| 0.13% | trace | 21% | **Oxygen** |
| trace | 0.0 | 0.0000017% | **Methane** |
| − 53 °C | + 459 °C | + 14 °C | **Surface temperature** |

Left: **Too cold, too hot, and just right for life: planetary temperatures and atmospheres.**

Is there, or at least was there, life on Mars? The debate began in 1895, when the astronomer Percival Lowell claimed to have observed 'canals' on Mars: '[his] vision of Mars as a desert planet, once Earth-like but now dying, gripped the 20th-century imagination', noted *Nature* in 2006. But the atmospheric composition and inhospitable surface temperature of Mars suggest that it is far more likely to be lifeless like Venus. However, the latest space mission to Mars detected traces of methane in the atmosphere.

Below: **Layered formations within an old impact crater in the Schiaparelli Basin of Mars, in a high-resolution image taken by the Mars Global Surveyor spacecraft in 2003. They immediately suggest comparisons with sedimentary structures on Earth, deposited at the bottom of ancient lakes or oceans and then subsequently exposed by geological weathering. 'The crater could well have been filled with water in the Red Planet's distant past, perhaps resting at the bottom of a lake filling the Schiaparelli impact basin', write Jerry Bonnell and Robert Nemiroff in *Astronomy: 365 Days*. But they concede that the layers might also have been formed by material settling out of the windy Martian atmosphere. The two Mars rovers, Spirit and Opportunity, in 2004 discovered intriguing hints of conditions suitable for Martian life in the past, such as unusual grey spherules, made of iron and rock, nicknamed 'blueberries', which may have been slowly deposited from some ancient bath of dirty water.**

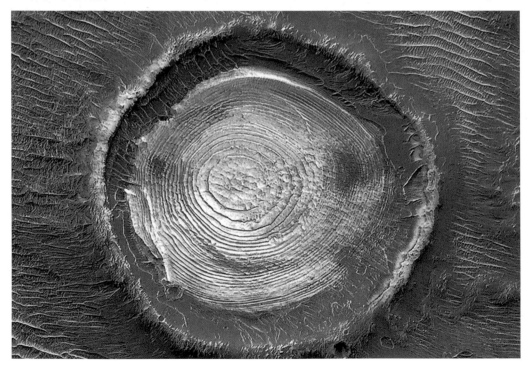

# The Sun

As the chief source of light and warmth, the Sun was central in all early civilizations, especially ancient Egypt, with its worship of Ra (as in Ra-meses). In Greece, the philosopher Diogenes, famous for living naked in a barrel near Athens, given the choice between riches and sunlight, chose sunlight. When Alexander the Great came one morning to test this claim by offering to grant Diogenes anything in the world he wished for, Alexander was allegedly told: 'I wish you wouldn't block the Sun.'

Reports of eclipses go back as far as Babylonian times in the 21st century BC. A cuneiform tablet records a prognostication that when the Moon god is in his eclipse, 'the king of Ur will be wronged by his son' – but that the son will not succeed to the throne, because the Sun god will catch him. Careful investigation suggests that this most probably refers to the murder of Shulgi by his son and the accession of Amar-Sin around the date of the lunar eclipse on 4 April 2094 BC.

The first blemishes on the Sun – the dark patches known as sunspots – were observed by Christoph Scheiner in 1611, using a telescope that projected the image onto a white screen, to protect his eyes. But Scheiner's early work had to be published pseudonymously to avoid bringing discredit on his Jesuit order, should it prove to be false. (This did not stop Galileo from identifying Scheiner and claiming he had already discovered sunspots.) In the 1620s, however, Scheiner published under his own name the first major work on sunspots, *Rosa*

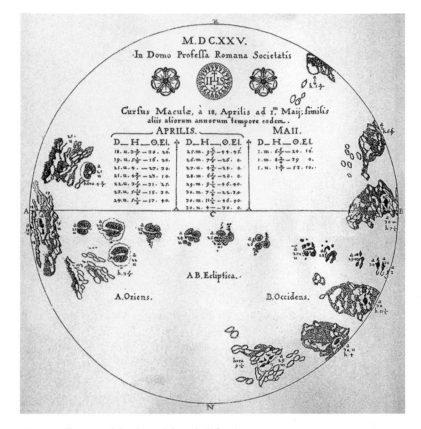

*Ursina*, illustrated by him (above). The photograph (below) shows an extreme close-up of a sunspot taken in 2002. It looks darker than the Sun's surface because it is more than 2,000 degrees cooler than the surface at a little under 6,000 °C.

Above: **Sunspots drawn by Christoph Scheiner, 1625.**

Below: **A close-up of a sunspot, 2002.**

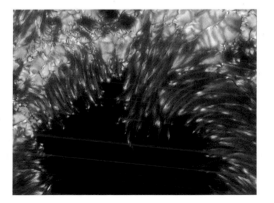

The astronomical term analemma was originally a Greek word, meaning the support or base of a sundial. It names the curve obtained when the position of the Sun is measured at the same time each day throughout the year. This task requires a camera positioned in exactly the same way for each exposure. The Earth-based analemma opposite, recorded in 2003 above ancient Nemea in Greece, required 44 separate exposures plus one foreground exposure. The Mars-based analemma below was simulated from photographs taken in 1997 from the Sagan Memorial Station established by the Mars Pathfinder mission. The tilt of Earth's axis and the variation in speed as the planet moves around its orbit produce the analemma curve. If, instead of being an ellipse, Earth's orbit around the Sun were a perfect circle, and if, instead of Earth's axis having a tilt, it were perpendicular to the plane of the orbit, the Sun would appear at the same point in the sky at the same time of day throughout the year – and the analemma would be a single dot. If the orbit were circular, but the polar axis tilted as it currently is, then the loops of the figure-of-eight would be equal in size.

**Analemmas on Earth** (opposite) **and Mars** (below). **The simulated Martian analemma is a teardrop, not a figure-of-eight like Earth's, due to a different interplay for Mars between the tilt of its axis and its orbital shape. Saturn, too, would have a teardrop-shaped analemma, while Jupiter would have an elliptical analemma.**

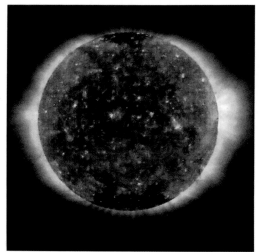

Early 1997

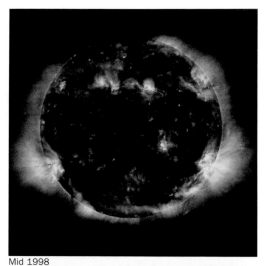

Mid 1998

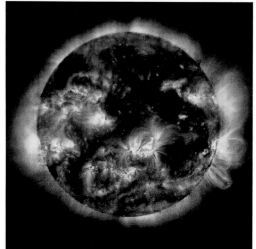

Late 1999

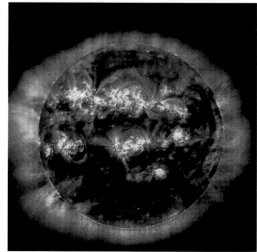

Early 2001

Left: **Cyclic solar activity. These ultraviolet images covering 1997–2001 show rising solar activity, peaking in 2001, during one of the Sun's 11-year cycles – as demonstrated by the graph** (below) **showing sunspot incidence over the period 1750–1997. 'Scientists generally agree that the energy released in a flare must first be stored in the Sun's magnetic fields',** comments Gordon Holman. **The process appears to involve 'magnetic reconnection', in which oppositely directed magnetic fields come together and partially annihilate each other – which explains the loops in solar flares.**

Bottom left: **Annular eclipse, 1994, in which the Moon blocks most, but not all, of the Sun's light. The Moon's orbit is not a pure circle, so at times it is slightly further away from the Earth, changing the geometry of the eclipse.**

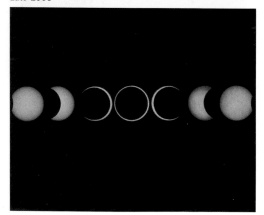

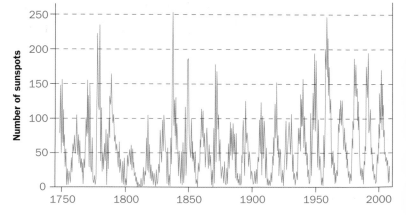

# Stars

The number of stars in the Universe runs into trillions – that is $10^{12}$ – only a very small percentage of which are visible to the naked eye. Although people have been gazing at the stars for thousands of years, and humanizing them as gods, animals and familiar objects like pairs of scales, the experience of looking up never fails to induce fresh awe.

The closest star to the solar system is Proxima Centauri, approximately 4.3 light-years from the Sun. (The light-year is a convenient non-SI unit equal to the distance travelled by light through a vacuum in one year; the SI equivalent is $9.461 \times 10^{15}$ metres.) The furthest stars lie billions of light-years away. Single stars like the Sun are less common than paired stars, multiple systems or clusters with numerous components. Stars have no uniformity in brightness, colour, temperature, mass, size, chemical composition or age.

Hipparchus, with his star catalogue of the 2nd century BC (see p. 22), was the first to introduce a system for measuring brightness. It had six levels of magnitude, with the 15 brightest stars of the northern celestial hemisphere in class 1 and the faintest stars in class 6. Today's system follows Hipparchus: the lower the magnitude, the brighter the star. For every decrease of 1 in the magnitude,

the brightness increases logarithmically by 2.512 times, which means that a decrease of 5 in magnitude corresponds to an increase in brightness of $2.512^5$, or 100 times. Some stars are so bright they have *negative* magnitudes, for example Sirius at –1.5; compare –4.0 for Venus, the evening and morning star, at its brightest, –12.7 for the full Moon and –26.9 for the Sun. These figures are all *apparent* magnitudes, since they do not take account of the distance of a star from Earth. Absolute magnitude, which directly measures a star's luminosity, is the magnitude that the star would have if viewed at a standard distance of 32.6 light-years.

The spectral analysis of radiation that has proved so powerful a tool in understanding the Sun (see p. 76) is the main technique for learning about the constitution of distant stars. They are thereby classified into spectral types, according to their colour and temperature, with the hottest blue type O having a temperature of 40,000 K and the coolest red type M of 3000 K. A graph of spectral type plotted against absolute magnitude yields the Hertzsprung-Russell (H-R) diagram, in which stars tend to cluster in certain parts, for instance the 'white dwarf' stars, the 'giant' stars and the 'Cepheid variables'.

Above: **Constellations, in a popular textbook of cosmography, 1540.**

Left: **A Chinese star chart, 1193, part of the longest continuous astronomical records of any civilization.**

Next page top: **Star trails above Mauna Kea, the location in Hawaii of very large optical telescopes such as the Keck telescopes (see p. 67). Over 150 one-minute exposures by a digital camera were used to make the photograph, during which time the rotation of Earth produced the impression of long star trails. The volcanic landscape in the foreground was illuminated by the Moon.**

Next page below: **The Andromeda Galaxy in infrared. The nearest major galaxy to our own Milky Way, this spiral galaxy is two million light-years away. The image was created from 11,000 separate exposures made over 18 hours by the Spitzer space telescope. Infrared light is particularly sensitive to dust heated up by stars.**

# Comets

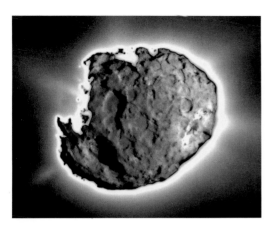

The visit of a comet has long fascinated mankind. Halley's comet, which returns to Earth every 76 years (most recently in 1986), has been sighted since 240 BC. When it appeared in 1066 at the time of the Norman Conquest of England, its image was embroidered into the Bayeux Tapestry. In 1910, Halley's comet was especially bright. Leonard Woolf (future husband of Virginia Woolf), then a government officer in a remote coastal district of tropical Ceylon, noted in his official diary how the local villagers regarded the comet as an evil portent (manifested in his strictness!).

Edmond Halley was the first to calculate a comet's orbit and forecast its return. In 1705, he announced that three comets observed in 1531, 1607 and 1682 were so similar in characteristics they must be the same comet; he predicted the next visit in 1758. When this duly occurred (after his death in 1742), the comet was named after him.

At the beginning of the 21st century, our knowledge has leapt ahead. Spacecraft have photographed comets close up, samples have been brought to Earth, and a probe has impacted a cometary nucleus. The major surprise is that 'these balls of dirty snow are born of fire as well as ice', observed *Nature* in 2006. Scientists were amazed to find a huge range of minerals in the particles captured by Nasa's Stardust probe as it swooped past the comet Wild 2 on 2 January 2004. 'Many of the compounds could only have formed close to a star – far from the chilly outskirts of the solar system where the comet first coalesced.'

Left: **Nucleus of comet Wild 2, recorded by the Stardust spacecraft during its fly-by in 2004. The nucleus was only about five kilometres in diameter, while the jets of dust and gas left trails millions of kilometres long.**

Below: **Halley's comet of 1066, as shown in the Bayeux Tapestry.**

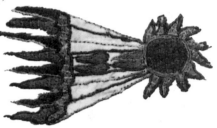

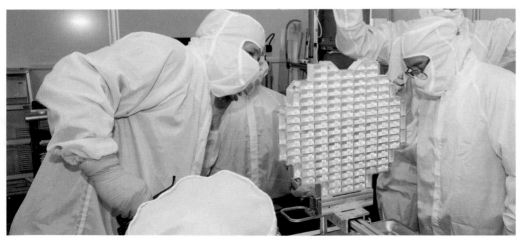

Left: **Scientists examine blocks of aerogel containing grains from comet Wild 2's tail. 'We could easily work for a year on a single particle.'**

# Black Holes

Given the hyperbole that surrounds black holes, and their leading proponent Stephen Hawking, it needs to be said that they may or may not exist – there is no absolute proof; also, that the idea is nothing new. It was proposed in 1783 by John Michell (the seismologist) and calculated by Pierre Simon Laplace soon after; defined mathematically by Karl Schwarzschild in 1916 using Einstein's general relativity; proved mathematically in 1939 by Robert Oppenheimer and Hartland Snyder but firmly repudiated by Einstein himself; and finally named 'black hole' by John Wheeler in 1964. Michell considered the possibility of an object large enough to have an escape velocity greater than the speed of light.

A modern definition, from the touring planetarium show 'Black Holes: the Other Side of Infinity' runs: 'A black hole is an extremely massive concentration of matter, created when the largest stars collapse at the end of their lives. Astronomers theorize that a point with infinite density – called a singularity – lies at the centre of black holes.'

Black holes, by definition, cannot be seen, because light cannot escape from them (though other radiation, dubbed Hawking radiation, can emerge). The only way to locate them is by their effects on the radiation from neighbouring objects in space, such as gas jets and rapidly orbiting bodies. The first good candidate for a black hole was Cygnus X-1, discovered in 1965.

The centre of the Milky Way. This near-infrared image, made by the Paranal observatory's Very Large Telescope, spans about two light-years of the galaxy. The orbit of a particular star, S2, may show the existence of a black hole at the centre of the Milky Way with over two million times the mass of the Sun. The claim is based on the orbit of S2 that appears to be dictated by enormous gravity exerted by an unseen object that must be extremely compact – a super-massive black hole. But some astronomers remain sceptical and prefer to postulate a new type of star to explain such movement.

# Constants of Nature

Even Einstein made a few mistakes in physics. By far the best known – because Einstein drew public attention to it – is his 1917 addition of the *cosmological constant* to his 1915 field equations of general relativity. This episode tells us something important about all physical constants.

Einstein added his cosmological 'fudge factor' only in order to make his proposed model of the Universe static and therefore eternal, which is how he and others thought the Universe should be. Without it, he argued, gravitational attraction must make the Universe fall in on itself. Since the constant was 'necessary only for the purpose of making a quasi-static distribution of matter', it seemed an ugly intrusion into the elegance of his original equations. Then Edwin Hubble's telescopic observations of galaxies proved that the Universe is expanding, not static. In 1931, a convinced Einstein told journalists he was withdrawing both his static model in favour of an expanding one (see p. 68) and his cosmological constant. Although Einstein much regretted having added it in the first place, he welcomed the increase in simplicity for his field equations.

'However the possibility of a cosmological constant did not go away so easily,' wrote a leading theorist and Nobel laureate, Steven Weinberg, in 1993 in *Dreams of a Final Theory: The Search for the Fundamental Laws of Nature*. 'It is not enough to say that a cosmological constant is an unnecessary complication. Simplicity, like everything else, must be explained.' Weinberg's caution was justified. Very recent observations of the expansion of the Universe strongly suggest that the cosmological constant does appear in the field equations after all. The reason appears to be connected with an acceleration in the expansion of the Universe.

There are many apparently fundamental physical constants. The one known to non-scientists is the speed of light, $c$, which is $2.977\ 924\ 58 \times 10^8$ metres per second (approximately 300,000 km/s). Others are the gravitational constant $G$ in Newton's equations of gravity, the charge on the electron or proton $e$, the Avogadro constant $N_A$ for the number of units in a mole (see p. 90), and the Planck constant $h$ that links the energy of a quantum with its frequency. (The energy is given simply by multiplying the frequency by the Planck constant.) Thousands upon thousands of experiments have shown that these numbers remain constant under very different physical conditions. If they were to have even slightly different values on Saturn from on Earth, for example, the calculations of the Cassini mission would have gone awry, and the spacecraft would have been lost in space.

Yet no theory explains why the fundamental constants have the values they have, or why they should stay the same and not vary. The constants are, in the final analysis, human constructs, like the laws of nature that have been simultaneously 'discovered' and 'invented' by scientists, from Archimedes onwards. The chequered history of Einstein's cosmological constant reminds us of this important truth.

**Albert Einstein (1879–1955), the most important physicist of the 20th century as a result of his foundational work in relativity and quantum theory. But Einstein accepted that his theories, like Newton's, would not be sufficient for all time. He wrote: 'Science is not and never will be a closed book. Every important advance brings new questions. Every development reveals, in the long run, new and deeper difficulties.' Einstein's proposed cosmological constant epitomizes the truth of his remark.**

# The Expanding Universe and the Big Bang

'Ten or twenty billion years ago, something happened – the Big Bang, the event that began our Universe. Why it happened is the greatest mystery we know. That it happened is reasonably clear.' Carl Sagan's statement sums up the present-day consensus among cosmologists, who favour a single moment of creation, currently thought to be some 13.7 billion years ago. To those who ask 'What came before the Big Bang?', the consensus maintains that 'time' did not exist 'before the Big Bang', and so this phrase is meaningless.

A moment of creation necessarily implies an expanding Universe – exploding from a point and receding in all directions. The first measurements providing solid evidence of expansion were Edwin Hubble's in the 1920s. After exhaustive, all-night observations with the world's largest telescope, the most sensitive photographic plates and an accurate spectrometer at the Mount Wilson observatory, Hubble was able to plot a celebrated graph in 1931 (see right). This shows how the recession velocity of galaxies varies with their distance from the Earth. The velocities were calculated from the Doppler shift (see p. 102) in the galaxies' wavelength, which gets longer – i.e. becomes shifted towards the red end of the visible spectrum – the faster the galaxy recedes. The distances of the galaxies from Earth were obtained from a clever technique developed by other astronomers based on comparing the apparent and absolute magnitudes (see p. 149) of the stars known as Cepheid variables, which show a periodic variation in their brightness.

Hubble's graph is a straight line, with due allowance for the uncertainty in the data. This meant, astonishingly, that recession velocity is proportional to distance. The further away the stars are, the faster they are receding. Indeed, Hubble calculated a constant of proportionality, such that the recession velocity is equal to the distance multiplied by the Hubble constant.

After Hubble's measurements, the most important evidence for the Big Bang comes from the cosmic microwave background (CMB) radiation. In the 1940s, George Gamow and others had predicted that the primordial light emitted just after the Big Bang should even now be detectable, flooding the Universe. They calculated the wavelength of its ripples, after allowing for the intervening expansion of the Universe by a factor of about 1,000 (based on Hubble's law), to be roughly 1 millimetre. This wavelength fell in the microwave region

**Edwin Hubble (1889–1953), the astronomer who discovered the expanding Universe in the 1920s. By measuring the recession velocity of galaxies, and their distance from Earth, he discovered that the velocity was proportional to the distance, as shown in his graph below – a relationship known as Hubble's law.**

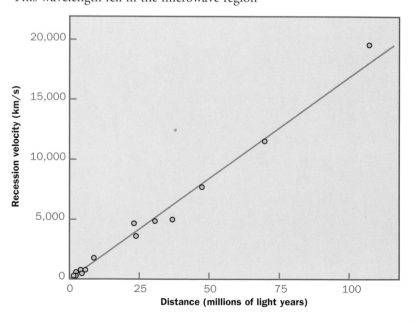

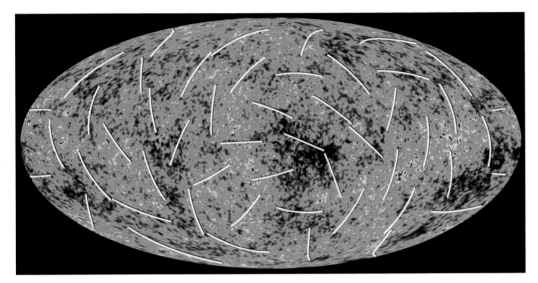

**Ripples from the Big Bang** (left), **as detected by the Wilkinson Microwave Anisotropy Probe** (below). **The WMAP, launched in 2001, measured the anisotropic cosmic microwave background (CMB) radiation with a resolution 35 times better than the Cosmic Background Explorer satellite of 1992. The different colours in the signal indicate tiny temperature fluctuations (between 30 and 70 millionths of a kelvin), while the white lines show different polarizations of the radiation. This polarization of the CMB appears to have occurred after the formation of the first stars, probably more than 13 billion years ago.**

of the spectrum. In 1965, such microwave radiation was detected, by chance, during the operation of a radio telescope by Arno Penzias and Robert Wilson. Fuller measurements from space, made in 1992 by satellite (the Cosmic Background Explorer), revealed that the CMB is actually not isotropic, i.e. the same in every direction, but *anisotropic*: it contains tiny fluctuations. These are ripples from the tiny fluctuations in density after the Big Bang: signals from the formation of the galaxies out of the isotropic primordial cosmic fog.

The Big Bang theory suffers, hardly surprisingly, from some formidable explanatory gaps. But it undoubtedly explains the available evidence from a wide range of astronomical observations better than the eternal Universe. The latter idea had developed, in the hands of a small group of important astronomers, from the 'static' Universe favoured by Einstein in 1917, so as to encompass Hubble's data on expansion.

It acquired a new name, the 'steady-state' Universe, in which expansion was allowed but new matter (i.e. new galaxies) was created to fill the gaps resulting from the expansion, leaving the Universe unchanged, with a steady overall density. How ironic, then, that the name Big Bang should have been coined as a wry joke by one of its adamant opponents, Fred Hoyle, who believed in the steady-state theory!

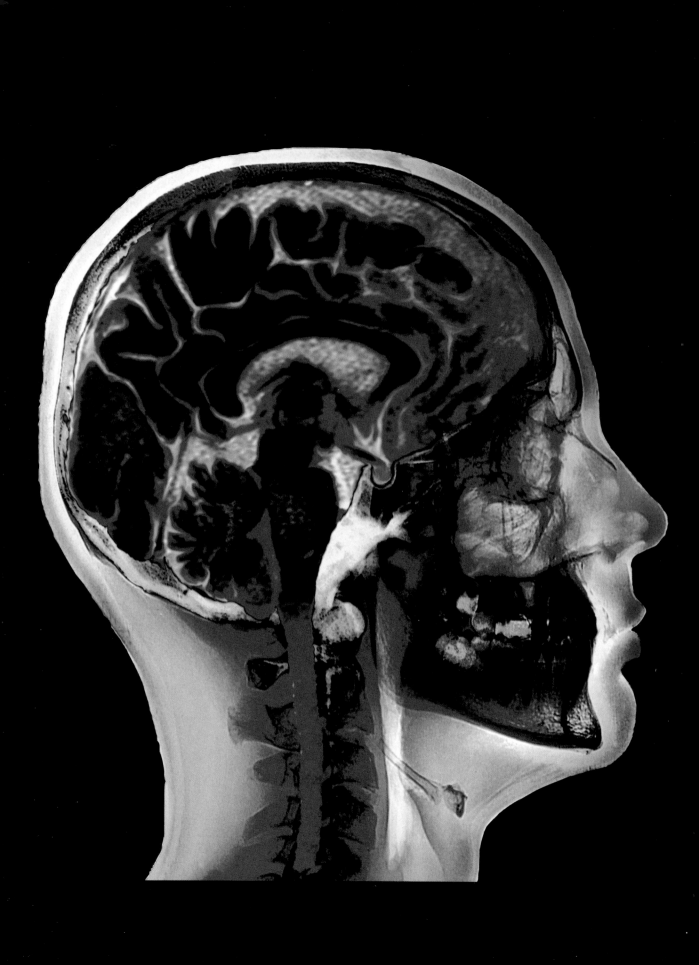

# III MEASURING MAN

'Men argue; Nature acts', said the pithy Voltaire. However, as Einstein sagely observed: 'The most incomprehensible fact about Nature is that it is comprehensible.'

Man and measurement may not always make for a happy marriage, yet the measurement of human nature – mind, body and society – increases decade after decade. Intelligence testing, educational and professional examinations and the subdivision of knowledge into specialisms, proliferate. So do health indicators, medical tests and surgical operations. At the social level, business is inconceivable without market research, government is impossible without tax and census returns, and national politics and elections are dependent on surveys and opinion polls.

Plainly, measuring human nature differs from measuring physical nature in the degree of precision and certainty achievable; and it is constrained by ethics that do not apply to experimenting on the planet or on subatomic particles. But the goal is still the same for a linguist, a psychologist, a geneticist or a sociologist, as for a physicist: to quantify a human phenomenon by careful and controlled observation, and so obtain data that can be analysed and theorized. We know that there are costs involved, but also that the benefits generally outweigh them.

Magnetic resonance imaging scan of a human head. MRI is invaluable for physicians and fascinating for psychologists. It shows the brain with amazing clarity, and may also reveal the mind.

Chapter 8 *Mind*

The Rosetta Stone, key to unlocking the Egyptian hieroglyphs. Discovered at Rosetta (Rashid) in Egypt in 1799 by French soldiers, it was soon taken to the British Museum as a spoil of war. It comprises three scripts – hieroglyphic (top), demotic (middle) and alphabetic (bottom). The hieroglyphic and demotic basically record the same Egyptian language, while the alphabetic script is in Greek. Today the Rosetta Stone has come to symbolize the power of writing and the mind over the material world.

# Language

During the mid-19th century, the American speech therapist Alexander Melville Bell devised a notation for recording speech, to help deaf students learn how to speak intelligibly. Its symbols looked nothing like the English alphabet but were abstract representations of the positions and movements of the articulatory organs of speech – tongue, lips, teeth and so on. For example, all symbols curving to the left represented consonants formed at the back of the tongue. Bell published his system in 1867 as *Visible Speech: The Science of Universal Alphabetics*. At public lectures he would call upon someone from the audience, preferably a person who spoke an unfamiliar dialect, to say a few words of his or her own choice, which Bell would write down. Then he would call his son, Alexander Graham Bell (later famous for his telephone), who had not heard the words, to read aloud from the transcription. The speaker's original pronunciation and the son's imitation would be compared.

What are the key differences between written and spoken language – apart from the self-evident fact that writing is visible and speech invisible? The most important is that a passage of writing naturally breaks down into its constituent symbols, be they letters of an alphabet, Chinese characters or Egyptian hieroglyphs, whereas a passage of speech does not. Of course we do often divide speech into consonants, vowels and syllables, and linguists create many other categories of spoken 'atoms and molecules', such as phonemes and morphemes. But these divisions are always artificial and never entirely free from overlap.

'Speech is a river of breath, bent into hisses and hums by the soft flesh of the mouth and throat', writes the linguistic scientist Steven Pinker in *The Language Instinct*. There are no spaces between words

Below: **The masthead of the phonetic journal *Le Maître Phonétique*, published in 1914; after 1970 the journal opted for a standard orthography, because readers found the fully phonetic orthography too demanding. The text reads: 'Le Maître Phonétique, organ de l'association phonétique international, vingt-neuvième année – janvier-février 1914'.**

Next page: **Sound, symbol and script. Ten scripts are used to write 'four score and seven years ago' (the opening words of Abraham Lincoln's Gettysburg Address). At the top is the acoustic spectrogram of the particular speaker. 1 shows the transcription in the International Phonetic Alphabet; 2 the English spelling; 3 the Russian alphabetic transcription; 4 the Bengali alphabetic version, with transcription; 5 the Korean Hangul version, with transcription; 6 the Egyptian hieroglyphic version (Ptolemaic period), with transcription; 7 the Arabic consonantal version, with transcription; 8 the Japanese syllabic version, with transcription; 9 the cuneiform syllabic version, with transcription; 10 the Chinese syllabic version, with Pinyin transcription.**

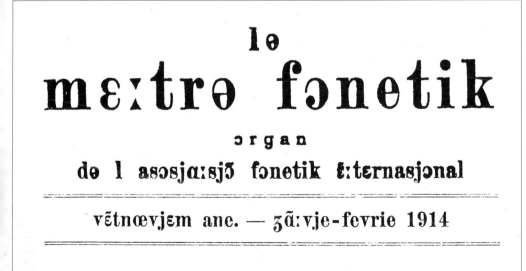

lə mɛːtrə fɔnetik
ɔrgan
də l asɔsjɑːsjɔ̃ fɔnetik ɛ̃ternasjonal
vɛ̃tnœvjɛm anc. — ʒɑ̃ːvje-fevrie 1914

in normal speech, as there are white spaces between words in most of today's writing systems. We may imagine that there are such gaps, but when we listen to speech in a foreign language, our delusion is exposed. Speech is a flow, constantly changing in frequencies, loudness and pitch. If we were to splice a tape of someone saying 'cat' into its two constituent consonants and one vowel (as near as can be done), and then reverse them, we would not hear 'tac' but something unintelligible. Over half the words we use in normal conversation we cannot recognize if they are replayed in isolation, because they are so rapidly and informally articulated.

Each spoken language has its own gamut of sounds, drawn from a literally infinite range of possible sounds. Its writing system represents some of this gamut – the phonetic proportion varies with the system – leaving readers to guess the rest. The divergence between sound and script is greatest with foreign words and names. Each of the ten scripts at right does the job of transcribing words differently and with differing degrees of accuracy. The International Phonetic Alphabet (IPA) at the top (1), introduced in 1888, is now so accurate that it represents even the accent of a speaker (as Bell's earlier notation endeavoured to do) – if in this example the speaker were British or French, instead of Chinese-American, the IPA transcription would differ. But the IPA's gain in phonetic accuracy is offset by its consequent lack of readability. All scripts must strike a compromise between accuracy to the mouth and intelligibility to the mind.

# Verse Metre and Scansion

*A little learning is a dang'rous thing;*
*Drink deep, or taste not the Pierian spring:*
*There shallow draughts intoxicate the brain,*
*And drinking largely sobers us again.*

Alexander Pope's words are still memorable mainly because they are true, but also because they scan. Pope's metre, highly popular in his time, is an iambic pentameter. Each line measures 5 feet – hence '*penta*meter' – the foot being the basic unit of verse, consisting of 2 or more syllables; Pope's foot has 2 syllables, making 10 syllables per line. Moreover, the syllables have a rhythm: the first is short/unstressed, the second long/stressed – a form known as an iamb, very common in English verse. Thus, the stresses lie on the first syllable of 'little', 'learning' and 'dang'rous', and on 'is' and 'thing'.

Most classic English poetry uses lines of 4, 5, 6 or 7 feet (tetrameter, pentameter, hexameter or heptameter). Apart from iambic, the foot can be: trochaic (2 syllables: stressed-unstressed), spondaic (2 syllables: stressed-stressed), anapaestic (3 syllables: unstressed-unstressed-stressed), dactylic (3 syllables: stressed, unstressed, unstressed) and amphibrachic (3 syllables: unstressed-stressed-unstressed).

In addition, forms can apply to a whole poem, such as the ballad, limerick and sonnet, not forgetting non-English forms like the haiku, rondeau and terza rima (Dante's form in his *Divine Comedy*). The Elizabethan or Shakespearian sonnet, for example, comprises 14 lines: 3 quatrains and 1 couplet, rhyming a b a b c d c d e f e f g g.

Some classic poems get away with breaking scansion rules but still sound as if they scan. For instance, Lewis Carroll's wonderful nonsense poem 'Jabberwocky' in *Through the Looking-Glass*, the first verse of which begins:

*'Twas brillig, and the slithy toves*
*Did gyre and gimble in the wabe;*
*All mimsy were the borogoves,*
*And the mome raths outgrabe.*

The first, second and third lines of each stanza seem to be in an iambic tetrameter, the fourth anapaestic. But who can be sure! For though the words sound meaningful, they are not to be found in the dictionary. As Humpty Dumpty explains to a puzzled Alice, 'toves are something like badgers, they're something like lizards, and they're something like corkscrews.' And: 'To gyre is to go round and round like a gyroscope. To gimble is to make holes like a gimlet.' Still puzzled? Below is the book's illustration of some toves gyring and gimbling.

Above: **Dante Alighieri (1265–1321). His epic poem, *The Divine Comedy*, written in Italian, uses iambic pentameter and *terza rima* (Italian for 'third rhyme'). This verse form consists of stanzas of three lines known as tercets. The first and third line of the tercet rhyme with one another, while the second line rhymes with the first and third lines of the following tercet. So the rhyme scheme of *terza rima* is a b a, b c b, c d c, … , y z y, z. This is a demanding scheme for languages less rich in rhyme than Italian, but from time to time it has been adopted by poets writing in English, such as Percy Bysshe Shelley and W. H. Auden.**

# Semaphore and Morse Code

The Roman Empire had a system of towers – more than 3,000 of them – for signalling warnings with lights. But not until the French Revolution did a code evolve for rapid long-distance communication. In 1794, Claude Chappe's 'aerial telegraph' began operating between Paris and Lille. Vocabulary words of a message, including names, could be encoded in the pre-agreed positions of a set of arms pivoted on a post, mounted on towers at distances of 8 to 16 kilometres – a semaphore (from the Greek meaning 'conveying a signal'). These positions were viewed from the next tower through telescopes. It sounds cumbersome, but with good visibility a signal could cover the 16 stations and 225 km to Lille in 2 minutes, and later the 116 stations to Toulon on the Mediterranean in 20 minutes. After refinement, Chappe's final code used 92 of the semaphore's possible 196 positions.

Around the same time, though independently, maritime signalling improved, such that a sailor holding small flags could semaphore messages to an observer with a telescope on another ship. The essential flag positions stood for alphabetic letters and numbers, unlike in Chappe's more complicated system.

The electromagnetic telegraph quickly supplanted semaphore on land. Morse code for sending telegraphic messages was invented in the 1840s by Samuel Morse. The operator pressed a key to send electrical pulses of varying lengths, creating a sequence of dot-and-dash combinations that represented letters and numbers. With Morse code, about a quarter more messages could be sent in a given time than with an arbitrary code.

Left: **The basics of the International Morse Code. The duration of a dash is equal to the duration of 3 dots. It is also the duration of the space that an operator must leave between each letter or other symbol; between words, a space of 5 or 6 dots is left. In devising the code, Morse's chief insight was that the briefest dot-and-dash combinations should represent the most frequently used English letters – which Morse discovered from the typecase of a Philadelphia newspaper. It contained 12,000 *e*'s, 9,000 *t*'s, 8,000 each of *a*, *o*, *n*, *i* and *s*, and so on; and so the code for *e* is simply 'dot', while that for *t* is just 'dash'. This idea accounts for the efficiency of Morse code. Nowadays, the International Morse Code is mainly of interest to radio hams. Their archetypal message is dash-dot-dash-dot, dash-dash-dot-dash, spoken as 'dahdidahdit, dahdahdidah'. It stands for CQ, which means 'Is anybody out there listening?'**

Bottom left: **Semaphore flag signals. In order to send numbers by semaphore, the signaller makes the 'numbers follow' signal, and then uses the letters of the alphabet for the numerals 1 to 9, for example, A = 1, B = 2, C = 3; the sign for J stands for 0.**

# Writing Systems

Purely pictographic writing, such as Ice Age cave art, and notations used today such as music and mathematics, cannot express any and all thought. Full writing – that represents the sounds of languages – developed through the *rebus* principle. This radical idea, from the Latin meaning 'by things', enables phonetic values to be represented by pictographic symbols. Thus in English, a picture of a bee with a picture of a tray might stand for 'betray', a bee with a figure 4 might represent 'before', while a picture of an ant next to a buzzing bee hive, might (less obviously) represent 'Anthony'. Egyptian hieroglyphs are full of rebuses, for instance the 'Sun' sign ⊙ pronounced *R(a)*

or *R(e)*, is the first symbol in the hieroglyphic spelling of the pharaoh Rameses. In an early Sumerian accounting tablet (see above) we find the abstract word, reimburse, represented by a picture of a reed, because 'reimburse' and 'reed' shared the same phonetic value *gi* in the Sumerian language.

Left: **A Sumerian rebus, c. 3000 BC.**

Below left: **Measurement of writing systems. This tree divides writing systems according to their nature, not according to their age; it does not show how one writing system may have given rise to another historically. (The dotted lines indicate possible influences of one system upon another.) How best to classify writing systems is a controversial matter. For instance, some scholars deny the existence of alphabets prior to the Greek alphabet, on the grounds that the Phoenician script marked only consonants, no vowels (like today's Arabic script). The root of the problem of classification is that there is no such thing as a 'pure' writing system – i.e. a system that expresses meaning entirely through alphabetic letters or entirely through syllabic signs or entirely through logograms (word signs) – because all full writing systems are a mixture of phonetic and semantic symbols. Nevertheless, labels are useful to remind us of the predominant nature of different systems.**

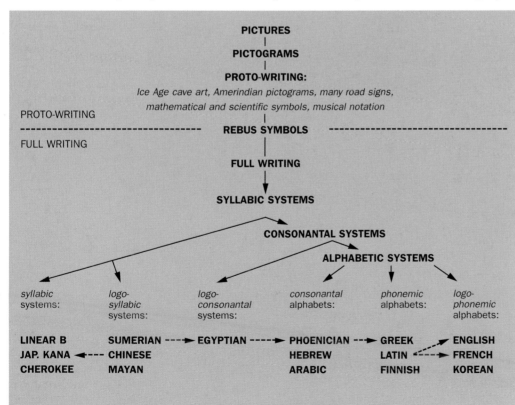

# Shorthand

The earliest shorthand was used by Xenophon to write the memoirs of Socrates. Since then hundreds of systems have been devised. Some use abbreviations of conventional spelling; others represent the sounds of speech; yet others require the learning of a list of arbitrary symbols; and some systems combine these different principles.

The best known was invented by Isaac Pitman in the 19th century. Its basic principle is phonetic, which has made it relatively easy to adapt for writing languages other than English. Some 65 letters are used, consisting of 25 single consonants, 24 double consonants and 16 vowel sounds. However most vowels are omitted, though they may be indicated by the positioning of a word above, on or below the line. The signs are a mixture of straight lines, curves, dots and dashes, as well as a contrast in positioning and shading. They relate to the sound system, for example, straight lines are used for all stop consonants (such as *p*). The thickness of a line indicates whether a sound is voiceless or voiced.

The shorthand used by Samuel Pepys for writing his famous diary was much less sophisticated. Invented by Thomas Shelton in the 1620s, in some ways it resembled an ancient writing system, such as Babylonian cuneiform. Although many of the signs were simply reduced forms of letters and abbreviations for words, there were nearly 300 invented symbols, mainly arbitrary logograms, such as 2 for 'to', a larger 2 for 'two', 5 for 'because', 6 for 'us'. (Several of these symbols were 'empty', presumably to foster secrecy.) Initial vowels were symbolized; medial vowels were indicated by placing the consonant following the vowel in five positions on, below or to the side of the preceding consonants; and final vowels were shown by dots, arranged similarly. Overall, the system was quasi-phonetic. Despite its drawbacks, it was popular in its day for reporting sermons and speeches, perhaps as fast as 100 words per minute.

**Samuel Pepys (1633–1703), diarist.** Below left **is the last page of Pepys's celebrated diary, which he gave up in 1669 when he was (mistakenly) convinced he was going blind, 'being not able to do it any longer, having done now so long as to undo my eyes almost every time I take a pen in my hand... And so I do betake myself to that course, which is almost as much as to see myself go into my grave: for which, and all the discomforts that will accompany my being blind, the good God prepare me!'**

Bottom left: **Pitman's shorthand.** Bottom right: **Speedwriting. 'Since the dawn of history man has strived to communicate with his fellows and to record experiences that would otherwise be forgotten.'**

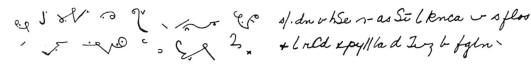

# Paper and Book Sizes

Ever since paper was invented in China, it has taken a wide variety of sizes. Most are obsolete as a result of the metric sizes introduced by Walter Porstmann in Germany in 1922 and now standard in almost all countries except the United States and Canada. These divide into three series – A, B and C – of which A is best known because of its use in technical drawings and posters (A0, A1), flip charts (A1, A2), drawings, diagrams, large tables and many photocopiers (A2, A3), letters, magazines, forms, catalogues and all photocopiers (A4), note pads (A5), postcards (A6) and even playing cards (A8).

Its advantage is obvious when one folds an A3 sheet to get two A4 sheets, or an A4 sheet to get two A5 sheets. (See diagrams.) The folding ratio works because A0, with an area of exactly 1 square metre, has sides in the proportion $\sqrt{2} : 1$ – as does every A paper size. This is the Lichtenberg ratio, named after the mathematician who noted its applicability to paper sizes in 1768 (not to be confused with the much older golden ratio).

Book sizes, following paper sizes, varied greatly, although the commonest sizes historically derived from a sheet of handmade paper measuring 19 × 25 inches. A printed sheet folded in half created the folio book (2 leaves and 4 pages), folded twice the quarto book (4 leaves and 8 pages), folded thrice the octavo book (8 leaves and 16 pages), and so on. Thus no paper was wasted, except in trimming. These size labels are still used, but they do not specify a book's actual dimensions unless the original sheet size is given.

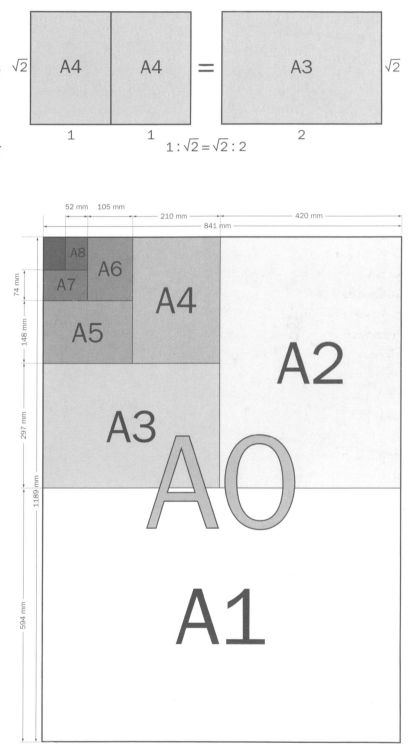

# Book and Library Classification

As knowledge expands and bifurcates, its classification in print must constantly change. Even a personal library needs rearranging periodically to reflect new areas of publication and its owner's shifting intellectual interests.

One of the earliest catalogues, compiled in Alexandria in the 3rd century BC, was known as *Pinakes* (Greek for 'tablets with wax in the middle'). Its entries were apparently arranged in at least ten, and maybe more, main classes, subdivided alphabetically by author. It became the model in the Middle East and Byzantine Empire until the early Middle Ages. However, medieval Europe's university libraries arranged books according to the trivium (grammar, rhetoric and logic) and the quadrivium (arithmetic, geometry, astronomy and music), the traditional subjects of study. All books had a fixed location, meaning that the librarian could never move them from shelf to shelf – a strange concept for the modern librarian used to a relative location.

Many library classification systems exist today, often associated with the world's great libraries, but the most used is the Dewey Decimal Classification (DDC) system, conceived by Melvil Dewey in the United States in 1876 and frequently updated since. DDC is mainly *hierarchical* and *enumerative*. Hierarchical, because it breaks subjects down into 'natural' subdivisions: first into 10 classes, such as Literature, then each class into 10 divisions, such as English and Old English Literature, then each division into 10

sections, for example English Fiction. Enumerative, because DDC assigns decimal numbers to all of these categories – here class 8, division 82, section 823. The DDC's 1,000 sections are numbered from 000 (Computer Science: Generalities: Knowledge) to 999 (Geography and History: General History of Other Areas: Extraterrestrial Worlds).

The chief alternative is a *facetted* classification system, increasingly attractive with the growth of the online world, where a classifying designation need not specify a shelf position. Facetted classification, notes Wynar's *Introduction to Cataloging and Classification* (9th edn), 'does not assign fixed slots to subjects in sequence, but uses clearly defined, mutually exclusive, and collectively exhaustive aspects, properties, or characteristics of a class or specific subject.'

In the Colon classification system – first used in the 1930s by S. R. Ranganathan – the five facets of the main subject are: 'personality' (the focal or most specific subject), 'matter', 'energy' (any activity, operation or process), 'space' and 'time'. Thus, a book on the eradication of virus in rice plants in Japan, published in 1971, is classified (using specified punctuation) as: J , 381 ; 4 : 5 . 42 ' N70 and facetted as follows:

| J | agriculture | (main subject) |
|---|---|---|
| 381 | rice plant | (personality) |
| 4 | virus disease | (matter) |
| 5 | eradication | (energy) |
| 42 | Japan | (space) |
| N70 | 1970s | (time) |

ISBN 1-85168-494-8

9 781851 684946

International Standard Book Number (ISBN). Introduced in 1969, the ISBN is now standard as a machine-readable code that identifies a book uniquely for all time, whether it is in print or not. Until 2007, ISBNs had 10 digits (ISBN-10); now, they have 13 digits (ISBN-13), beginning with 978 specifying that the number refers to a book. The other 10 digits identify group, publisher and title. The final 10th digit – the check digit – is a clever way of checking that an ISBN has been correctly given. Consider the above ISBN-10: 1 85168 494 8. Take the first 9 digits and multiply them by the numbers 1–9, as follows: $1 \times 1 = 1$; $8 \times 2 = 16$, $5 \times 3 = 15$, etc. Then add up these 9 products (in this case 250), divide it by 11 and you get 22, remainder 8. The ISBN is correct if the remainder matches the 10th (check) digit – here 8. (If the remainder is 10, the check digit is X.) The ISBN-13, here 9 781851 684946, is the same as the ISBN-10, except for the 978 prefix and the check digit, here 6, which is calculated in a different, slightly more complex way than for the ISBN-10.

# Typography

When Gutenberg invented printing with movable metal type and published his Bible in the 1450s, his typeface was black letter, based on the thick-stroked script known as Old English or Gothic. Black letter became the standard typeface in Germany. But during the 20th century, especially under the Nazi regime, black letter was the subject of major debate: supported by Adolf Hitler as patriotic in feeling, it did not appeal to Joseph Goebbels, Hitler's propaganda minister, who used the roman and the modern sans-serif typefaces in publicizing the 1936 Berlin Olympics. Then, in early 1941, Hitler made an about-turn, banned black letter as being a Jewish innovation, and decreed the standard should be roman.

Roman type, of course, has its origins in the capital letter forms of the ancient Romans, which were established by the 1st century AD; the lower-case roman letters were modelled on a version of the standardized script imposed by Charlemagne in the 8th century, known as the Carolingian minuscules. It also gave birth to the fourth major western typeface apart from black letter, sans-serif and roman: italic. When chancery scribes in Spain and Italy wrote the roman script at speed, it became italic.

From the 16th century, typographers created new type founts in roman and italic, and later sans-serif. Some typographers are commemorated in the names of typefaces still in regular use, for example, Claude Garamond, William Caslon, Giambattista Bodoni, John Baskerville and Eric Gill. Today, with personal computers, anyone can try their hand at designing type, by playing with its four characteristic features: sizing, x-height (see diagram), set (or width) and fit (the spacing of combinations of letters). The almost infinite possible permutations both please and assault our eyes wherever we look.

**Ascender** The parts of the strokes of the lower-case letters that project above the x-height

**x-height** The distance between the baseline and the top of the lower-case x

**Baseline** The line on which the letters 'sit'

**Descender** The parts of the strokes of the lower-case letters that hang below the x-height

Above: **Times New Roman, a ubiquitous typeface, originally designed in 1932 for The Times newspaper by Stanley Morison, who said of his only typeface design: 'It has the merit of not looking as if it had been designed by somebody in particular.' The type size here is 72 points, which is the height of the capital letters and approximately the distance from the top of a lower-case ascender to the bottom of a descender (this relationship varies with different letters and different typefaces). There are almost exactly 72 points in 1 inch; in the UK and US, the point is 0.351 mm, and in Europe 0.376 mm. Leading – that is, the space between lines of type – is also measured in points.**

Left and below: **Black letter, the typeface used in Gutenberg's Bible.**

# Photography

Ever since photography's invention in the 1830s by Louis Daguerre (using metal plates) and Henry Talbot (using paper), it has uneasily combined science with art. One of its pioneers, John Herschel, introduced the name photography, taken from two Greek words together meaning 'drawing with light', as well as the terms negative and positive, and snapshot. But Herschel was primarily an astronomer and physicist, and saw photography much more as a science than an art (though he encouraged the artistic photography of his close friend Julia Margaret Cameron). 'Herschel preferred drawing's contemplative observation to photography's snapshot', admits *The Oxford Companion to the Photograph*. Henri Cartier-Bresson, one of the greatest of photographic artists, took a view similar to Herschel's. Cartier-Bresson abandoned photography in his sixties and took up drawing and painting in earnest; he hung no photographs on the walls of his apartment in Paris.

Measurement is integral to photography. Focal lengths of camera lenses – telephoto, macro, wide angle or zoom – determine the distance at which objects can be clearly seen and the field of view. Aperture controls the amount of light passing through the lens, and hence its depth of field: the range over which the image appears sharp. The shutter sets the duration of exposure of film emulsion or digital charge-coupled device. Other important measurements deal with recording sensitivity and image quality, such as film speed, contrast (of lens, film, enlarger and paper), density and resolving power.

A scientific art. Multiple photographic image of the surface of Mars and the Viking lander, taken by an official of the space agency Nasa, 1976.

Modern cameras have obviated the need for amateurs to worry about most settings – but professionals must understand them.

Digital cameras, which first came on the market around 1990, have recently taken over from film cameras among the large majority of ordinary photographers, but the changeover is not a foregone conclusion among professionals. The resolution of top-of-the-range digital images is roughly equivalent to that of 35 mm film images – around 12 megapixels – and they are unquestionably more convenient to compose, process and store. But the resolution of medium-format and large-format films is still much finer than that of digital photographs. Moreover, random noise in digital cameras makes them unsuitable for long exposures; film is unaffected by this problem. Finally, film and prints maintain their quality, if stored under ideal conditions, for more than 100 years, whereas digital storage media currently degrade much more quickly and also employ formats, such as JPEG, that may not be easily readable in decades to come.

# Computing

A century ago, there were already 'computers' – but they were people, not machines. Human computers crunched scientific data too, although their speed of computation was very slow. Theirs was necessary, if tedious work, especially in astronomy, but far indeed from the almost magical powers we associate with today's electronic computers.

The internet, a network of computers capable of communicating with each other – and with it electronic mail – arose in the 1970s, originally from military research on how to protect command-and-control weapons systems located around the United States from nuclear attack. The key idea, as shown in the diagram, was to have a *distributed*, not a centralized network, so that the computer at each military station was connected to every other station via a number of links, not through a single link to a central headquarters vulnerable to attack. In addition, the network was given a deliberate amount of redundancy – in other words, more links than were strictly necessary for communication. (Language and scripts have plenty of redundancy too, to aid spoken and written communication, which shorthand largely eliminates – see p. 164.)

But to make this network function in practice, the transmission of electrical signals via the links had to be improved. With conventional analogue signals – continuously varying electromagnetic waves – of the kind used in telephone lines, the ratio of useful signal to useless noise decreased with distance, and the signal was soon corrupted by noise. However, corruption was not so problematic with digital signals, created by regularly sampling the continuous analogue signal and converting the samples into discrete numerical values, and then turning those numbers into sequences of bits (binary digits consisting of either 1 and 0 – see p. 38). A digital signal, being discrete rather than continuous, could have its transmission errors detected and corrected fairly easily. A so-called parity bit, rather like the check digit in the ISBN (p. 166), can be added to a digital signal at the end of each group of digits during transmission, and checked by the receiving computer to see if the parity bit corresponds with the number of 1's (an odd or an even number) detected in the received signal. If it does not match, then the receiving computer automatically requests retransmission of the group. 'It's the kind of thing computers can do a million times a second without batting a virtual eye', comments John Naughton in *A Brief History of the Future*.

Above: **Pixellated photograph of Alan Turing (1912–54), the 'father' of modern computing. His 1937 theoretical paper, 'On computable numbers', is his most famous mathematical contribution, which was followed by his invention of the concept of programming a computer. He wrote: 'The engineering problem of producing various machines for various jobs is replaced by the office work of "programming" the universal machines to do these jobs.'**

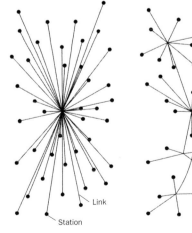

Centralized network

Decentralized network

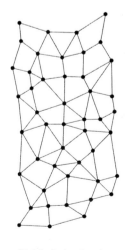

Distributed network

# Tools, Nails and Screws

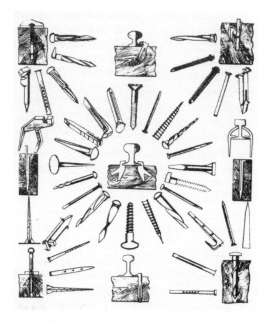

The earliest artifacts of humanity are generally thought to be tools, such as carved bones and flint axes. In the science-fiction film *2001: A Space Odyssey*, the discovery of the first tool at the dawn of man is shown unforgettably, when an ape idly picks up a heavy animal bone and uses it to bash first animal prey and then ape enemies to death. Triumphantly, the ape hurls the bone into the air, it spins over and over, and in a blink of the camera's eye, leaping an eon of evolution, the bone is transformed into a spinning space station.

'Few classes of artifacts exhibit as much diversity and specialization of form as the tools of the crafts and trades,' notes the engineer Henry Petroski in *The Evolution of Useful Things*. But the purpose of many surviving tools is lost. Tool-makers and tool-users were usually illiterate, and they deliberately kept tools secret for fear of competition. When a stranger entered their workshop, artisans would stow their tools away. If any questions were asked about them, the workmen might give frivolous or totally misleading answers.

However, there are some obvious long-lived categories like the hammer, axe/chisel, knife, drill and saw. The first saws were probably the jawbones and teeth of dead animals. The discovery of copper in the Near East some 4,000 years ago led to metal saws, made from successively harder metals: bronze, iron and finally steel. In the West, the cutting action generally occurred on the push stroke, whereas Oriental saws cut on the pull stroke. Among the more important developments were crosscut saw teeth, like knives, for cutting wood across the grain, and ripsaw teeth, like chisels, for cutting it along the grain.

| | |
|---|---|
| ⊨▭▭▭▭────────▷ | Round wire nail |
| ⓪▭▭▭▭────────▷ | Oval wire nail |
| ◖▭▭▭─────────▷ | Lost head nail |
| ◖──────────── | Cut clasp nail |
| ⊨•••••───────◁ | Clout nail |
| ⊢───────────◁ | Panel pin |
| ⊨─── ⊢──── | Tack |
| ◖──────── | Chair nail |
| ▯───────────▭ | Masonry nail |
| ▮⊞⊞⊞⊞⊞⊞⊞────▷ | Annular nail |
| ◂───────── | Hardboard pin |
| ◖∿∿∿∿∿∿─── | Galvanized nail |

Left: **Nails of the past. From the time of the Romans, and perhaps before, until the end of the 18th century, nails were made by hand (by 'nailers'). In the 19th century, machine-made 'cut nails' almost totally replaced hand-made nails. This illustration from the late 1800s, shows the nails available in the United States. It was compiled from nail patents by Benjamin Butterworth, then commissioner of patents.**

Below left: **Some, but by no means all, of the current varieties of nail, many of which have specialized purposes. The serrations on the shank of many nails are designed to increase friction with wood and prevent the nail from pulling out. In Europe nails are sold by length and diameter in metric units, but in the US the length is designated by penny size. This appears to be derived from an old English custom of selling nails by the hundred. 'Eight penny' nails would cost 8 pennies per 100 nails, for example. Since the English abbreviation for penny was 'd' (from the Roman *denarius*), US nail sizes are given in the form 3d, 4d, 6d and so on. The smallest size, 3d, is 1.25 inches long and the largest size, 60d, measures 6 inches.**

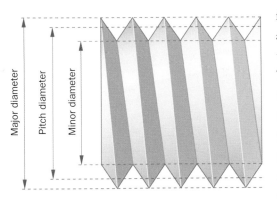

Major diameter | Pitch diameter | Minor diameter

The screw was allegedly invented by Archimedes for pumping out ships. A helical screw like a drill bit was enclosed in a cylinder at an angle; when the screw was rotated with a handle, its end scooped up water and the helix channelled this to the top of the screw. The screw principle was later used in olive and wine presses.

As fasteners, metal screws and nuts came only in the 16th century, when they were turned with a box wrench or sometimes a pronged device; the hand-held screwdriver dates from around 1800. In 1841, Joseph Whitworth designed the first unified screw-thread system, named after him, and in 1884, the BA system, named after the British Association for the Advancement of Science, followed, for screws of very small diameter used in scientific instruments and fine mechanisms. Even so, when tests were done around 1900 on 1,207 different combinations of nuts and bolts supposedly of the same size, poor design allowed only eight per cent to 'engage sufficiently' to be screwed up with a spanner!

Most screws and bolts are tightened as far as reasonably possible. However,

for critical applications, such as the nuts on a car's wheel, a specific torque is applied for tightening (settable on power wrenches). The idea is to stretch the bolt and compress the material between the head and nut, so that each is tensioned like a spring. The stretch is called a preload. If an external force then tries to separate the bolt, there will be no strain, unless the force exceeds the preload. The preload is calculated as a percentage of whichever is the weakest of the following: the bolt's yield tensile strength, the strength of the threads it screws into, or the compressive strength of the clamped material.

Some screws are self-tapping: their sharp threads create their own holes in materials such as wood, plastic and soft metal (below left). Others require a pre-drilled or pre-tapped hole, or a nut with a thread that matches that of the screw. The screw's major diameter (its 'size') is the outer diameter of the thread, which is bigger than the diameter of the hole or nut by an amount equal to the pitch of the thread (see diagram). Almost all threads are right-handed, so that one tightens a screw clockwise, but the reverse is true for left-hand threads. These are used in only a few situations: for instance, where the rotation of a shaft would gradually loosen a right-handed nut, as on a left-side bicycle pedal; in some gas-supply connections to prevent dangerous misconnections; and in the common device known as a bottlescrew or turnbuckle (below). Bottlescrews can adjust tension in ropes, cables and tie rods. They work by having a right-hand screw matched with a left-hand screw. Tightening the first screw also tightens the second screw, and the same is true of loosening.

# Music and Singing

Sound intensity has a physical magnitude independent of any listener, measured in decibels, whereas loudness depends on the perception of an individual listener (see pp. 101–102). Something similar is true in music of frequency and pitch. The frequency of a musical note can be measured in hertz, but its pitch, though governed by its frequency, is to a considerable extent a subjective measurement of 'highness' or 'lowness'.

Consider the eight notes of the octave in the western musical scale. Over this important interval, the frequency of a note doubles (or halves): if a piano's middle C is tuned to 261.6 Hz, its C above middle C must be tuned to twice this frequency, 523.3 Hz. But the pitch relationship between these two notes is one 'to which our ears are so attuned that we hardly think of it as a musical interval; rather it seems a kind of identity or semi-unison between the higher and the lower pitch', writes Arthur Klein in *The World of Measurements*. What affects us is the ratio of the frequencies more than their absolute values. Thus, in parallel with the eight notes of the octave, composers speak of *tones* in the full chromatic scale (eight basic tones, plus five sharp and flat semitones), and also pitch intervals based on the tones such as a minor third, a perfect fifth and a major seventh. Their values cannot be scientifically defined, unlike their frequency, but tone relationships are what makes music melodious or discordant.

Within the octave, the frequency ratios between the various notes and the sizes of the pitch intervals can be expressed either in decimal terms or with a logarithmic scale divided into 1,200 cents, as in the table below, which shows the generally accepted 'tempered' scale of tuning, favoured by J. S. Bach. The decimal ratio of the notes C above middle C to middle C is 2.0, as expected, and the logarithmic ratio is 1,200 cents. But the equivalent ratios for G above middle C (392.0 Hz), which has a frequency half way between middle C (261.6 Hz) and C above middle C (523.3 Hz), are 1.498 and 700 cents – not 600 cents, half of 1200. The reason for this is the logarithmic nature of the cent scale. Its great advantage over the decimal ratio scale is that each pitch interval is equal to 100 cents. This means that by subtracting or adding cent values, a composer can determine the musical intervals between any two tones.

Opposite: **The physics of singing. The difference between a trained singer and an untrained student is shown by acoustic spectra such as this one. The top part displays the fundamental frequency against time of a C-major arpeggio on the vowel in 'bee', as sung by a trained singer** (left) **and an untrained singer** (right); **the bottom part reveals all the harmonics (higher-frequency resonances) that give colour and power to a singing voice. The trained singer moves more smoothly through the notes in the arpeggio, and has powerful harmonics in the 2.5–4 kHz region.**

| Musical Scale of Equal Temperament | | | | | |
|---|---|---|---|---|---|
| Tone no. | Key | Pitch interval | Decimal ratio | Ratio (cents) | Frequency (Hz) |
| 1 | C | Unison | 1.0 | 0 | 261.6 |
| | C#, D♭ | Semitone/ minor second | 1.059 | 100 | 277.2 |
| 2 | D | Whole tone/ major second | 1.122 | 200 | 293.7 |
| | D#, E♭ | Minor third | 1.182 | 300 | 309.3 |
| 3 | E | Major third | 1.260 | 400 | 329.6 |
| 4 | F | Perfect fourth | 1.335 | 500 | 349.2 |
| | F#, G♭ | Augmented fourth/diminished fifth | 1.414 | 600 | 370.0 |
| 5 | G | Perfect fifth | 1.498 | 700 | 392.0 |
| | G#, A♭ | Minor sixth | 1.587 | 800 | 415.3 |
| 6 | A | Major sixth | 1.682 | 900 | 440.0 |
| | A#, B♭ | Minor seventh | 1.782 | 1,000 | 466.2 |
| 7 | B | Major seventh | 1.888 | 1,100 | 493.9 |
| 8 | C | Octave | 2.0 | 1,200 | 523.3 |

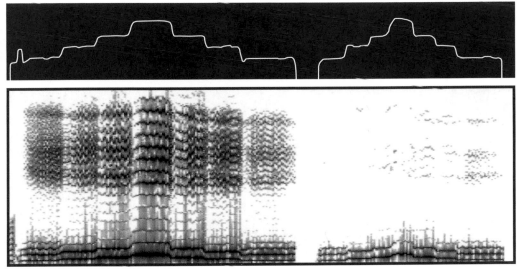

Below: **Autograph by W. A. Mozart. The opening of the Serenade for strings in G major (K. 525), which Mozart described as 'Eine kleine Nachtmusik', completed in Vienna on 10 August 1787, while he was writing his opera *Don Giovanni*. The time signature is denoted by the 'C' at the beginning of the stave, meaning 'common time', that is 4/4 time. 'This might, quite simply, be considered the most beautiful piece of occasional music ever written, hence its enduring appeal' (H. C. Robbins Landon in *Mozart: The Golden Years*).**

Notes can change frequency – unlike their frequency ratios. What is today known as concert pitch or standard pitch is officially taken to be 440.0 Hz for the note A above middle C. But in northern Europe from about 1500 to about 1670, the frequency of A was 466 Hz (a semitone above concert pitch), and from about 1670 to 1770, it was 415 Hz (a semitone below).

In musical staff notation, pitch is expressed by a note's level on the five-line musical stave. The modern stave generally uses two clefs indicating the pitch of the notes written on it: the treble clef and the bass clef. In the treble clef, the five lines of the stave from bottom to top represent the notes E, G, B, D, F – from a third above middle C to an octave-and-a-fourth above middle C. The four spaces between the lines therefore represent the notes in between: F, A, C and E. A note's duration is indicated by a series of symbols, such as crotchets ♩ (quarter notes) and quavers ♪ (eighth notes).

Reading across, the stave is also divided by vertical bars, determined by the time signature (rhythm) of the music; for example, 4/4 time means 4 beats to the bar, each of the duration of a crotchet or its equivalent, with the first beat often (though not necessarily) stressed.

# IQ and Intelligence

Intelligence testing is an extremely controversial field, partly because of its original association with eugenics and its continuing taint of racism, but mainly because no one, especially psychologists, can agree on a definition of intelligence. Everyone agrees that Einstein was highly intelligent, but what about a leader like Gandhi, an artist like Mozart or a sportsman like Pelé? They, too, obviously had extraordinary mental powers, but are these best described as 'intelligence'?

The British psychologist Charles Spearman and the American psychologist Lewis Terman, who independently laid the groundwork for intelligence testing in the early 20th century, believed that intelligence could be measured by a single number – $g$ for 'general intelligence' (Spearman) and IQ for 'intelligence quotient' (Terman).

In 1912, Terman designed the Stanford-Binet test (Stanford was his university and Binet the name of the French educationist whose test Terman adapted), which became the standard intelligence test for fifty years in the United States for schools, companies, the military and government. In 1917, a version was used to grade raw army recruits for the First World War. Illiterates took the purely pictorial Army Beta test. For example, they had to complete rapidly the missing element in each of the above illustrations. Plainly, any illiterate unfamiliar with American urban life was bound to score low (especially on item 15, which shows bowling). But this did not stop intelligence tests from acquiring an unwarranted reputation.

Above: **Part of the Army Beta intelligence test, 1917. (See text for explanation.)**

IQ was calculated by Terman in the following way. Each test item was given a 'mental age', at which a child should know the correct answer. If a 5-year old answered a test item graded for a mental age of 6 correctly, the child received extra points, with more extra points for a correct answer to an item graded for a mental age of 7, and so on. By totting up the total number of test points, a mental age in years and months was calculated. The IQ tester then divided the mental age by the child's chronological age and multiplied by 100 to avoid decimals. An IQ of exactly 100 therefore indicated that mental age equalled chronological age.

For decades, Terman monitored the progress of a group of children in California with high IQs of 135 or more, the so-called 'Termites', and their offspring, to investigate whether high IQ predicted high ability in

Opposite: **MRI scans of brain (right and left side). Red indicates a strong positive correlation between cortical thickness and IQ, purple a strong negative correlation.**

later life and whether IQ was inherited. By conventional standards, the Termites did very well as adults, becoming doctors, lawyers, businessmen and scientists at leading American institutions at a rate vastly above that expected from the general population; and their children had high IQs too – less than 20 per cent were below 120. But no Termite was awarded a Nobel prize; moreover, Terman's initial tests rejected two future Nobel-winning scientists (Luis Alvarez and William Shockley). None founded an industry. And none became a well-known leader, artist or athlete.

Even so, the concept of 'general intelligence' continues to fascinate psychologists, not to mention the public. The arrival of brain scanning by MRI (see p. 182) inevitably suggested trying to link intelligence with the development of the brain. A group of American scientists at the National Institute of Mental Health tracked a group of more than 300 children with IQ tests and brain scans as they aged from 6 to 19, and published the results in *Nature* in 2006. These show a negative correlation between higher IQ and thickness of the cerebral cortex, particularly in the frontal and temporal regions, in young childhood, which reverses in late childhood and becomes a positive correlation for most of the cerebral cortex. In other words, the cortex is *thinner* in higher-IQ young children than in the average-IQ young child. The researchers conclude: '"Brainy" children are not cleverer solely by virtue of having more or less grey matter at any one age. Rather, intelligence is

related to dynamic properties of cortical maturation.' In saying this, the scientists of course continue to make the assumption that IQ tests measure intelligence.

A completely different, and somewhat less controversial measure of intelligence is to count up how often an academic's published works are cited by other academics in their writings. The more frequently an academic is cited, the more influential, and therefore in a certain sense the more intelligent, the academic is. But quite apart from an obvious flaw in this argument – that ten people can be just as wrong as one – the measure takes no account of who cited the work or where the citation was published. To be cited by a Nobel laureate in a sought-after journal is likely to be a better indicator of intelligence than to be cited by an obscure academic in a journal read by few. To adjust the bare statistics of citation, other indices, such as a journal's 'impact factor' and the popularity of the academic citing the work, as measured by Google's PageRank algorithm, have therefore become significant. But, for all this effort to measure intelligence, one must not forget those undoubted geniuses who chose not to write down their ideas, such as Socrates!

Above: **The Royal Society, London. Election to a fellowship is a major intellectual honour, but of course it is influenced by politics. A new index, the 'h-index', suggested by an American physicist, Jorge Hirsch, attempts to make elections to such bodies fairer. It is the highest number of papers a scientist has published that have each received at least that number of citations. So an h-index of 50 means that someone has written 50 papers that have each had at least 50 citations. The aim is to recognize not merely high productivity of papers or frequent citation of a few papers, but a large and consistent body of significant work.**

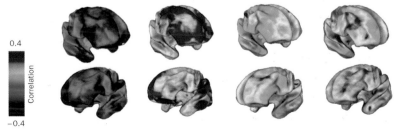

0.4
Correlation
−0.4

Young childhood    Late childhood    Early adolescence    Early adulthood

*Proportional Study of a Man in the Manner of Vitruvius*, pen and ink drawing by Leonardo da Vinci, *c.* 1487. This famous drawing expresses the classical ideal of the Roman architect Vitruvius, shared by many Renaissance artists including Leonardo, that architecture should be based on the divinely ordered proportions of the human being. It shows how both the circle – centred on the navel with the circumference defined by the outstretched arms and legs – and the square, can be derived from the proportions of a man placed flat on his back.

# The Human Genome

No one doubts the seminal role of genes in controlling the body's development, yet paradoxically, the more detail geneticists gather about the human genome, the harder it is to define the gene. 'Try defining the word gene – you will not find it easy', wrote Francis Crick in 1992, some 40 years after he and James Watson discovered the structure of DNA (deoxyribonucleic acid), the basic genetic building block, from which enzymes, proteins and chromosomes are synthesized.

Crick's perception was confirmed in 2006 by a couple of 'experiments' on different groups of geneticists. In one, researchers in science philosophy sent fourteen different sets of real genetic information to 500 biologists and asked them to complete a questionnaire and decide whether each set represented one, or more than one, gene. In the other experiment, twenty-five gene sequencers were closeted together by a colleague and asked to come up with a definition of 'gene' all could work with.

In the first case, the researchers found that 60 per cent of the biologists were typically sure of one answer, and 40 per cent confident of another. Almost none confessed to ignorance. In the second case, after arguing heatedly for nearly two days, the best that the twenty-five scientists could manage was a loose definition of the gene as: 'A locatable region of

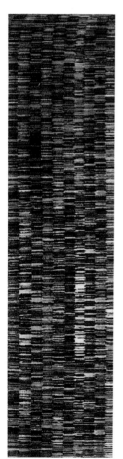

genomic sequence, corresponding to a unit of inheritance, which is associated with regulatory regions, transcribed regions and/or other functional sequence regions.'

In an editorial on this confusion, *Nature*, the scientific journal that originally published the structure of DNA in 1953, admitted: 'Decades of discussion have left a rather widespread perception, embraced by the general public and the media, of the gene as a tightly defined entity that spells out an inescapable destiny filled with beauty and health or, more often, blemishes and disease.' While this picture was clearly inadequate, said the journal, the erosion of the simple mechanistic idea that there should be a gene 'for' this particular disease and another gene 'for' that type of behaviour, was surely welcome. 'The genetic code holds new allure – its 4-letter sequence may have been documented but it contains deeper hidden ciphers, and geneticists relish the task of breaking them.'

Perhaps, rather than code breaking, we should instead expect the future progress of genetics to be more like foreign-language learning. We start with vocabulary lists, dictionaries and word-for-word decipherment and one day, after much effort and exposure to native speakers, we are able to appreciate the ambiguity and richness of great literature.

Above: **The double helix of DNA, the basic genetic building block. The framework of the helix is composed of sugar-phosphate units; the rungs are formed by the four nitrogenous bases adenine (A), guanine (G), thymine (T) and cytosine (C).**
Left: **In the output from an automated DNA sequencing machine, each colour represents one of the four bases. DNA (then known as nuclein) was chemically isolated in 1869; shown to be the substance of inheritance in 1944; and structurally analysed in 1953.**

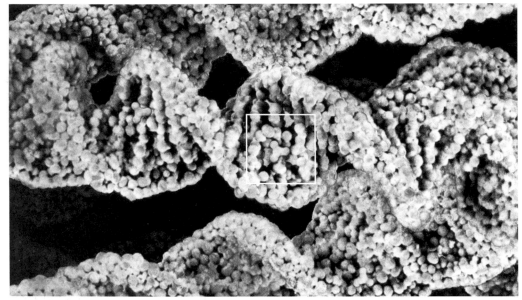

A close-up of the double helix of DNA. The scale line represents 10 nanometres ($10^{-8}$ m). The genome for a particular organism depends on the running sequence of the differing rungs of the molecular ladder, composed of the four differing bases A, G, T and C (see previous page). During cell duplication, the two rails of the ladder separate, and each acts as a template for making a complete new copy of the ladder of rungs.

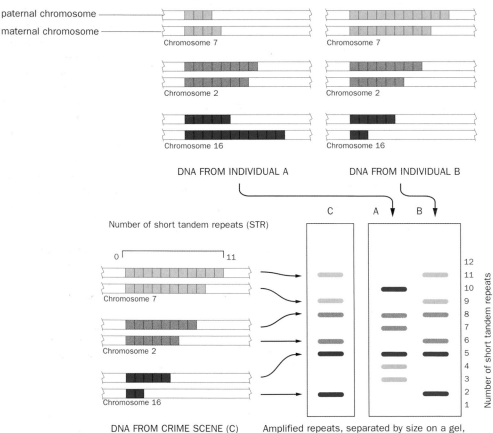

paternal chromosome
maternal chromosome

Chromosome 7          Chromosome 7
Chromosome 2          Chromosome 2
Chromosome 16         Chromosome 16

DNA FROM INDIVIDUAL A          DNA FROM INDIVIDUAL B

Number of short tandem repeats (STR)

0          11

Chromosome 7
Chromosome 2
Chromosome 16

C     A     B

12
11
10
9
8
7
6
5
4
3
2
1

Number of short tandem repeats

DNA FROM CRIME SCENE (C)

Amplified repeats, separated by size on a gel, give a 'DNA fingerprint'

Genetic fingerprinting. In order to match DNA taken from a crime scene with the DNA of potential suspects beyond legal dispute, short tandem repeats (STRs) are compared. STRs are sequences of two to four bases that recur as many as 17 times in DNA. The diagram shows that the STRs in the DNA from the crime scene C match those in the DNA of individual B but not those in the DNA of individual A. Whether police forces should store the genetic data from DNA samples of past suspects for possible future checks is controversial. Alec Jeffreys, the geneticist who invented DNA fingerprinting in the mid-1980s, has described such archiving as 'a gross infringement of civil liberties.'

# Blood

William Harvey, the physician of King Charles I who discovered the circulation of the blood around 1628, believed that blood was 'the fountain of life and the seat of the soul'. The five litres of blood in a typical human body (the individual's capacity depends on body weight at the rate of about 70 ml/kg), has always had an emotional significance far beyond its physical function – as witness the Christian sacrament.

Before Harvey, the beliefs of Galen, the most famous physician of ancient Rome, prevailed: blood was produced continually in the liver and sent around the body, seeping through the heart, by a sort of boiling generated by heat from the digestion of food. Harvey used measurement and calculation to disprove this. He measured the capacity of the heart in human beings, in dogs and in sheep. Then he multiplied this figure by the pulse rate. This allowed him to compute the amount of blood transferred from the heart in a given time – approximately 80 pounds (36 kilograms) in each half hour for an average man. His calculation was inaccurate, but it nevertheless showed, beyond doubt, that 'the beating of the heart is continuously driving through that organ more blood than the ingested food can supply, or than all the veins together at any given time can contain.' Hence blood was being constantly circulated rather than purely replenished.

The understanding of blood types – which makes successful blood transfusions possible – took until the 20th century; they were the discovery of the immunologist Karl Landsteiner and two colleagues, beginning in 1900. The four types identified were later given the labels we use today: A, B, O and AB. Type O is the oldest and was the only blood type in the Stone Age; A appeared around 25,000–15,000 BC; B emerged between 15,000 and 10,000 BC; and AB is the newest type, somewhere between 1,000 and 500 years old. Partly for these reasons, the frequency of blood types varies around the world. In the United Kingdom, 42 per cent of the population is of blood type A, 10 per cent of type B, 44 per cent of type O and a mere 4 per cent of type AB. In the United States, Central and South America, most people are of type O. Type A is common in Central and Eastern Europe and is the commonest group in Norway, Denmark, Austria, Armenia and Japan. Around a quarter of the people in Chinese and Asian communities are of type B. As for AB, the rarest type, around 10 per cent of the population of Japan, China and Pakistan have this type.

Blood type is inherited, with one 'blood type' gene – O, A or B – coming from the mother and a second one from the father. The two blood type genes together determine a person's blood group. The possible gene combinations are therefore OO, AA, AO, BB, BO and AB. Crucially, however, the A and B genes are 'dominant' genes and trump the O gene – so the AO combination belongs to blood type A, the BO to type B, and only the OO to type O. This explains why two parents of blood type A, for example, can produce a child of type O – both parents must have had the gene combination AO.

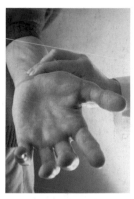

Feeling the pulse. The most common place to feel a pulse is where the radial artery lies near the surface on the thumb side of the wrist. A healthy adult has a pulse rate of 60–70 beats per minute when resting, but the rate can be as low as 40 in a top-class long-distance swimmer and as high as 80. The less elastic the arteries are, the more sharply the pulse pressure rises with each heartbeat. Every minute, the left ventricle of the heart pumps typically five litres (all the blood in the body) into the aorta and arteries. With each beat, the arterial pressure comes to a peak, known as the systolic pressure, and a minimum, the diastolic pressure. These are measured using a sphygmomanometer (an inflatable cuff connected to a mercury manometer) and a stethoscope. Thus, a blood pressure of 120/80, signifies a systolic pressure of 120 mmHg and a diastolic pressure of 80 mmHg.

Also important in classifying blood is the rhesus factor: Rh positive or Rh negative. In 1940, Landsteiner and colleagues injected blood from a rhesus monkey into rabbits and guinea pigs. The antibodies produced reacted with the red cells of some humans ('Rh positive') but not with those of others ('Rh negative'). The cause was found to be the presence or absence of an antigen known as D in a person's cells; those having the D antigen are Rh positive, and their cells are attacked by D antibodies in the blood of a person who lacks D and is Rh negative. Most humans are Rh positive (83 per cent in the UK population), so the blood bag below is relatively uncommon.

**Aplastic anaemia**
up to 2½ litres (about 4 pints) a month
● ● ◖

**Hip replacement surgery**
up to 3 litres (about 5 pints)
● ● ●

**Cancer**
up to 4½ litres (about 8 pints) a week
● ● ● ● ◖

**Brain surgery**
up to 5½ litres (about 10 pints)
● ● ● ● ● ◖

**Heart surgery**
up to 14 litres (about 25 pints)
● ● ● ● ● ● ●
● ● ● ● ● ● ●

**Car crash or gunshot victims**
up to 28 litres (about 50 pints)
● ● ● ●
● ● ● ● ● ● ●
● ● ● ● ● ● ●
● ● ● ● ● ● ●

**Liver transplant**
up to 57 litres (about 100 pints)
● ● ● ● ● ● ●
● ● ● ● ● ● ●
● ● ● ● ● ● ●
● ● ● ● ● ● ●
● ● ● ● ● ● ●

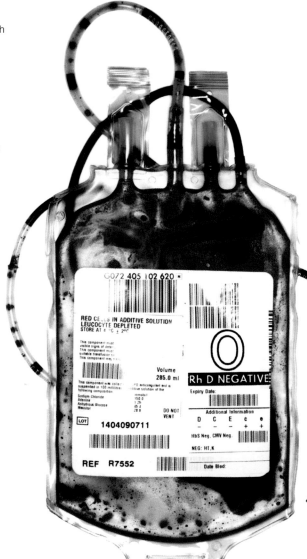

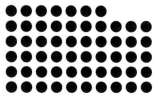

Blood transfusion. Both the blood type (A, B, AB or O) and the rhesus type (positive or negative) must be matched in a transfusion. Otherwise, there is the risk of haemolysis: the rupture of the recipient's red blood cells leading to the loss of haemoglobin, the oxygen-carrying molecule in blood. Haemolysis happens when the recipient's blood plasma contains antibodies to the donor's red blood cells, which attack the latter. Blood type A has anti-B antibodies in its plasma, and blood type B has anti-A antibodies, while type AB has neither of these antibodies, and type O has both the antibodies. From their murky beginnings in the mid-17th century, transfusions have become commonplace, to replace blood lost through accident or surgery or low haemoglobin levels as a result of illness. More than 40 million pints of donated blood are used every year in the US, Europe and Japan; the diagram shows the differing amounts used in a range of operations. Before being given to patients, the various constituents are separated – chiefly into red blood cells, platelets (cell fragments that promote blood clotting), and plasma. The blood bag in the photograph contains 285 ml (half a pint) of red cells belonging to blood type O, Rh D negative.

# Medical Scanning

Nineteenth-century physicians generally resisted instruments in diagnosis. The stethoscope was acceptable because only the physician could hear it, but devices with a read-out, like a thermometer, or worse still, with a written trace, were a potential threat to public confidence in expert medical judgement. 'They stood for an ethic of impersonal facts rather than personal trust, and were long held suspect by most doctors in America and Europe', comments Theodore Porter in *Trust in Numbers*.

The greatest advance in medicine in the 20th century, along with synthetic drugs, came from new, non-invasive instrumental techniques able to create astonishing images of bones, tissues, organs and body systems. The five main types of scan are now X-rays, computed tomography (CT), ultrasound, nuclear medicine and – the most recent development – magnetic resonance imaging (MRI). Each has strengths and weaknesses.

CT uses a high dose of X-rays to obtain a series of adjacent cross-sections known as tomographs, from which a computer creates high-resolution, three-dimensional images of internal organs – impossible with conventional X-ray images. Ultrasound uses pulses of very high frequency sound, and has no known hazards, so it is ideal for imaging

The main medical imaging techniques compared. Highlighting indicates that a technique is the preferred one for a particular area of the body.

| | X-RAYS | CT | ULTRASOUND | NUCLEAR | MRI |
|---|---|---|---|---|---|
| Bone | still the most common modality, giving the best resolution | used for more complicated structures, such as skull | poor – ultrasound will not penetrate bone | good for early diagnosis (e.g. of stress fractures), and whole body bone cancer, especially detection of metastases | not good for bones as gives weak MRI signal, but excellent for joints and preferred modality for knees – dynamic information can be obtained |
| Brain and spinal cord | radiograph of limited use | excellent for bone fine structure, competes with MRI for soft tissue diseases | poor – difficult to image through the skull without surgery | good for degenerative disease and receptor sites | excellent – the preferred modality for strokes and tracking of nerve fibres |
| Chest | radiograph gives adequate routine lung screening | CT preferred for better detail | poor – ultrasound cannot image past air spaces | very good for functional studies of both air and blood flow | little used – MRI not good for imaging air spaces |
| Heart and circulation | of little use | excellent with multislice scanners – 3D volume acquisition to improve temporal resolution in progress | excellent – the preferred technique in many cases. Velocities (Doppler), as well as structure analysed | useful for blood flow and function such as uptake of fatty acids | excellent with very good delineation – improved temporal resolution in progress using tracers for blood flow |
| Soft tissues (joints) | radiograph gives poor contrast | good and preferred to MRI for extra bone details | reasonable, although bone blocks ultrasound | resolution poor, but gives functional information | excellent – the preferred method for studying muscle, tendons, cartilage, joint cavity |
| Soft tissues (abdomen) | radiograph poor and needs contrast medium | good for many types of disease but contrast required – virtual colonoscopy becoming widely used | excellent – by far the preferred choice in obstetrics, since safe and 3D imaging now available | important functional tests for tumours, liver, kidneys, sentinel nodes, etc. – poor resolution but diagnostically valuable | widely used for tumours, after injection of various contrast materials |
| Comfort and safety | radiograph gives small radiation dose | high dose | no known hazards | moderate dose due to administered radionuclides | pacemakers, implants etc. a hazard. Some claustrophobia |
| Examination time | very fast | moderate, i.e. a few seconds | moderate | a few minutes or more, waiting for tracer distribution | long |
| Spatial resolution | 0.1 mm | 0.25 mm | 1–5 mm | 5–15 mm | 0.3–1 mm |
| Mobility | small portable machines available | none | portable machines widely used | portable devices developing | very limited, despite attempts at development |

unborn babies. In nuclear medicine, a radioactive tracer swallowed by the patient is monitored and reveals how circulatory systems such as blood and air are *functioning* – unlike X-rays. MRI employs an incredibly strong magnetic field. Since the nuclei of many body atoms behave like tiny bar magnets, they align with the field, though not perfectly. By making the field oscillate at different radio frequencies, and pulsing it, MRI scanners can cause different nuclei to resonate and absorb energy. When the pulse ends, the resonance ceases and the nuclei re-emit the energy – as a weak MRI signal – and this is detected by receiver coils.

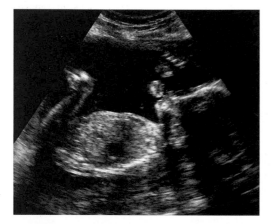

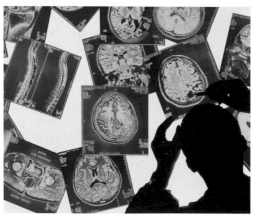

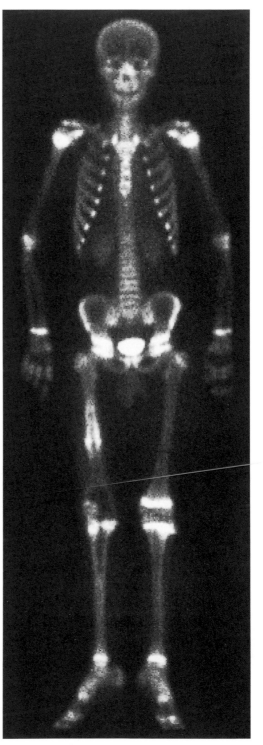

Left: **A whole body scan. It uses a radioactive tracer 'attached' to a convenient chemical compound. It shows a young teenager, in whose bones the tracer has accumulated. Rapid growth produces typical increased activity in the joints, but in this case not in one of the knee joints. The cause is a tumour in the leg above the knee.**

Far left: **An ultrasound scan showing an unborn baby. Such scans can identify heart, brain or spine abnormalities, locate the placenta and forecast the delivery date.**

Bottom left: **MRI scans give extraordinarily clear images of brains and soft tissue. They detect brain activity by discriminating between the magnetic resonance of oxygenated and non-oxygenated haemoglobin molecules. Yet it is by no means clear how this blood-oxygen-level-dependent MRI signal relates to neural activity, despite bold claims that MRI can show links between particular parts of the brain and particular types of behaviour.**

# Eyes and Lenses

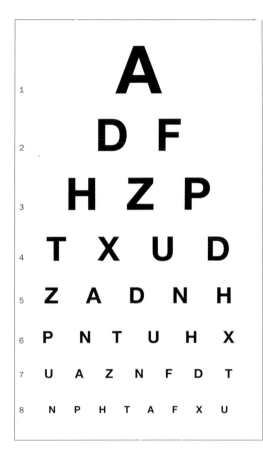

Nero, wrote Pliny, viewed gladiatorial games through the transparent green gemstone beryl, in order to correct his eyesight; and a glass bowl full of water was used as a magnifier in antiquity. But the first spectacles did not emerge until around 1280 in Venice, the centre of glass-making. Opinion was initially divided. A sermon in Pisa in 1305 regarded 'the art of making glasses' as 'the most useful of arts that the world possesses.' But an English vicar declared: 'The newly invented optick glasses are immoral, since they pervert the natural sight and make things appear in an unnatural and a false light.'

By the time Newton published his *Opticks* in 1704, the laws of refraction by lenses, both converging and diverging, and by prisms, were empirically understood, leading to improved spectacles, telescopes and microscopes during the 18th century. Light itself was not understood, but scientists could calculate the bending of light.

The eye and vision remained a mystery, too. How did the eye accommodate itself to objects at different distances, bringing each into focus? What kinds of distortion did it suffer from? How did it discriminate between different colours? And what was the relation between the eye and the mind? Answers to such questions – and practical ways of measuring and correcting the eye's flaws – began to be given around 1800 by Thomas Young, and they are still being given today in techniques such as laser refractive surgery for those who wish to avoid wearing spectacles or contact lenses, and in the many rival theories of conscious experience. As Leonardo da Vinci recognized: 'The eye, which is the window of the soul, is the primary means by which the brain may most fully and magnificently contemplate the infinite works of nature…'

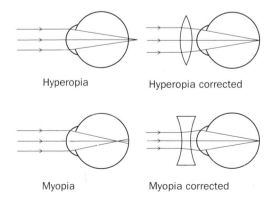

Hyperopia      Hyperopia corrected

Myopia      Myopia corrected

The Snellen chart (left) is named after Hermann Snellen, who developed it in 1862. Viewed from 6 metres, or 20 feet, its 8 lines of optotypes measure visual acuity: a person's ability to discriminate form. With 'normal' vision, acuity 1, a person can read line 7 – also known as 6/6 vision in the metric system or 20/20 vision in the US. Air force pilots must have unaided vision of the last line, that is 6/5 vision (acuity 1.2); UK car drivers 6/10 vision (acuity 0.6), which falls somewhere between lines 5 and 6. Snellen's notation means, very roughly, that with 6/12 vision (20/40 in the US) a person must approach to a distance of 6 metres to read letters that with 6/6 vision are readable at 12 metres. Poor visual acuity is correctable with convex lenses for hyperopia (long sight) and concave lenses for myopia (short sight). A further lens correction is often required for astigmatism, discovered by Thomas Young (above, 1773–1829).

# Body Mass Index

The body mass index (BMI) has been around for a fairly long time, since 1830–50, when it was invented by Adolphe Quetelet. But only in the past few decades has it become the most familiar way of classifying human weight from the point of view of health. A BMI in the range 18.5–25 is regarded by many physicians and medical organizations as indicating 'healthy weight'. Less than 18.5 is 'underweight', 25–30 is 'overweight' and more than 30 is 'obese'.

According to this measure, over half of the adult population of the United Kingdom is overweight and one fifth is obese; the equivalent figures for the United States are two thirds overweight, and one third obese. Globally, about a billion people in a population of 6.5 billion weigh in with a BMI of more than 25. On the other hand, there is no doubt that BMI as a health indicator takes insufficient account of physical fitness. Since muscle weighs more than fat, healthy athletes will often have a high BMI, despite their being very fit. The index is also biased against tall people.

Obesity is unquestionably rising in the rich nations. However, BMI may not be the most effective indicator of its health implications. Waist-to-hip ratio – with a limit of 1.0 (men) and 0.9 (women), as recommended by the British Heart Foundation – may be a better indicator of risk from heart attack in most ethnic groups.

Far left: **BMI is calculated by dividing weight (in kilograms) by the square of height (in metres). So someone who weighs 75 kg and is 1.8 m tall has a BMI of $75/(1.8)^2$, which equals just over 23, that is in the 'healthy weight' category. The categories are: <18.5 = underweight, 18.5–25 = healthy weight, 25–30 = overweight, >30 = obese. The height and weight ranges for each BMI category appear opposite. But BMI on its own is at best a rough-and-ready indicator of risk from disease, not a diagnostic test.**

Left: **How old are you, really? To understand the true connection between body weight, health and mortality, scientists will have to discover more about how different parts of the body age and replace themselves – or not, as the case may be. This diagram shows the average age of cells in different organs and tissues, where known. Some parts, such as the gut epithelium, skin and red blood cells, are quickly replaced, while brain cells are as old as our chronological age and, disturbingly for us, never replace themselves. (Experimental claims to the contrary have so far proved unrepeatable.) The average age of fat is one of the unknowns currently under investigation.**

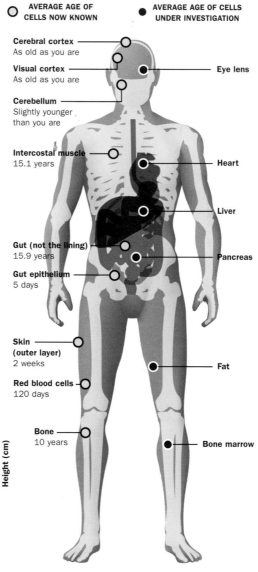

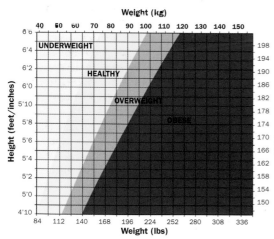

# Calorie Count and Food Additives

The amount of food energy we need per unit of body weight falls as we age. A new-born infant needs three times as much per kilogram as a 25-year-old woman, and a 10-year-old boy twice as much per kilogram as a 65-year-old man. On the other hand, humans, not being very active creatures, consume food equal to only 1–2 per cent of their body weight daily – as compared with a hummingbird or a small mouse, which must eat more than its own body weight each day.

To calculate how much food energy you need, first calculate how much your body at rest requires for its metabolism, just to maintain life. Then calculate the energy consumed by your various activities during the day – as indicated in a list such as the one on the right. Add the two energies together and the total is the daily amount needed from food for equilibrium, with no weight loss or gain.

On average, for every kilogram a person weighs, the energy required for metabolism is 1 kilocalorie per hour. The kilocalorie is the unit of food energy, confusingly equal to the Calorie (with a capital C) and loosely called the 'calorie' in conversation and on some food packaging, equivalent to about 4.2 kilojoules.

So a person weighing 70 kg burns 70 × 24 = 1,680 kcal per day while simply ticking over. If he or she spends the entire day (16 hours) sitting and reading, an extra 16 × 80 = 1,280 kcal will be needed, totalling 2,960 kcal per day. This is slightly more than the guideline daily amount of food energy, 2,500 kcal, recommended for men (2,000 kcal for women). If by contrast the person spends 16 hours playing golf (clearly unrealistic), the total will be 6,480 kcal, or in digging the garden, 10,000 kcal. These calculations take no account of the loss of energy (around 10 per cent) in the body's processing of food for storage and use, known as the thermic effect. But it is obvious that the guideline amount is not a satisfactory general nutritional recommendation on its own; it needs to be supplemented by knowledge of the energy required by one's typical daily activities.

| Activity | Energy consumed per hour by a 70 kg adult (kcal) |
| --- | --- |
| Sitting | 80 |
| Sweeping the floor | 120 |
| Croquet | 160 |
| Driving a car | 192 |
| Playing table tennis | 200 |
| Sailing | 200 |
| Walking slowly | 220 |
| Ironing | 240 |
| Cycling slowly | 260 |
| Polishing the floor | 272 |
| Canoeing and rowing | 280 |
| Playing golf | 300 |
| Walking fast | 320 |
| Doing ballet | 320 |
| Surfing | 320 |
| Skating | 360 |
| Gentle jogging | 360 |
| Basketball | 360 |
| Gymnastics | 400 |
| Climbing | 440 |
| Playing tennis | 480 |
| Jogging | 480 |
| Digging the garden | 520 |
| Playing football | 560 |
| Cycling fast | 672 |
| Skiing downhill | 700 |
| Hard running | 800 |
| Playing squash | 920 |
| Swimming fast | 1,020 |
| Climbing uphill when skiing | 1,120 |

Ratatouille Niçoise à l'huile d'olive
Ingrédients :
Légumes 83% (courgettes, tomates, aubergines, oignons, poivrons rouges et verts), jus de tomate, huile d'olive (1,5%), sucre, sel, amidon modifié de maïs, arômes naturels (dérivés de céleri, dérivés de lait), jus de citron, acidifiant : acide citrique (E330), épaississants : gomme xanthane (E415), gomme guar (E412), épice.

POUR 100 G DE PRODUIT :
VALEURS ENERGETIQUES MOYENNES
kcal 26,1 - kJ 110
VALEURS NUTRITIONNELLES MOYENNES
PROTEINES 1,0g - GLUCIDES 3,5g
LIPIDES 0,9g
CONTENANCE  POIDS NET TOTAL
425 ml  375 g

E numbers. E stands for Europe and refers to a European Union systematization of food additives, including natural vitamins, in a shorthand form suitable for the limited space on many food labels. The E system is based on the International Numbering System (INS) determined by the Codex Alimentarius committee, a body established in 1963 by the UN's Food and Agriculture Organization and the World Health Organization. Only INS additives approved by the EU receive an E number. The main categories are: E100–199 (colours), E200–299 (preservatives), E300–399 (antioxidants, acidity regulators), E400–499 (thickeners, stabilizers, emulsifiers), E500-599 (pH regulators and anti-caking agents) and E600–699 (flavour enhancers). For example, the antioxidant ascorbic acid, vitamin C, is E300, and the flavour enhancer monosodium glutamate (MSG) is E621.

# Alcohol Content

Aleconners, once the official inspectors of ale, were one of the most extraordinary group of professional measurers in recorded history. Indeed, aleconners seem to have behaved more like characters from *Monty Python's Flying Circus* than respectable local government officers. They would be summoned when an alehouse keeper had a batch of new brew: he would hang a bush out of his first-floor window (hence the common use of 'Bush' in the names of English pubs). Thus invited, the conners would enter the establishment, wearing the leather shorts or shortened trousers that were part of their measuring equipment. The ale to be assessed would be poured onto a bench and when it had formed a thin film, an aleconner would sit on it. After waiting a few moments, he would try to get up. If this was easy, he would declare the ale fit for the penny price, but if he tended to stick to the bench, the ale, being of a higher specific gravity, would be judged worth the higher rate, two pence. Then, presumably, the officer would be invited to sample the ale and would go away to wash his trousers!

The traditional British test of distilled alcohol is much better known. An excise officer would mix a sample of the spirit with a small amount of gunpowder and try to ignite the combination. If it burnt briskly, it was declared to be 'above proof'; a slow burn corresponded to 100 proof, that is the proof state; and if it failed to ignite, it was obviously below proof. The test was by no means foolproof or precise, and was later replaced by specific gravity (density)

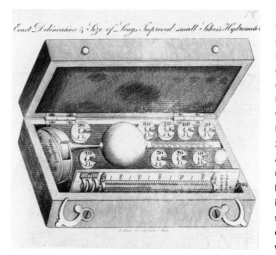

*Exact Delineation & Size of Songs Improved small Sikes's Hydrometer*

measurement using a hydrometer. In the metric system of measuring alcohol (named after its inventor, the French chemist Joseph-Louis Gay-Lussac), based on percentage by volume, 100 proof is equal to 57 per cent alcohol, pure alcohol is about 175 proof, and most brandy is around 70 proof, 40 per cent alcohol – which is why it will not catch fire on Christmas puddings.

Left and below: **Hydrometers for measuring the strength of alcoholic beverages. In 1802, the UK Board of Excise held a competition to design a better hydrometer, and it was won by an employee, Bartholomew Sikes. From 1816, the Sikes Hydrometer Act set the legal standard until 1907, and the Sikes hydrometer remained in use until 1980. The standard version** (left) **had one float and ten brass weights, and a mercury thermometer. Floated in the spirits or wort** (below, middle image), **the Sikes hydrometer was weighted until the zero mark on its graduated stem was level with the liquid surface. Tables were provided to convert the weight added into degrees proof, while also taking account of temperature.**

# Air Quality and Pollen Count

London's Great Smog of 1952 virtually halted road and rail transport and cost the lives of around 4,000 people. Such flagrant pollution may be history in Europe and the US, but air pollution most certainly is not – whether in the West or in the mega-cities of India and China such as Mumbai and Beijing. Today's air contains other types of pollutants, many from vehicle exhausts. They are less visible than coal-fire and factory smoke, but equally toxic, causing eye irritation, asthma and bronchial complaints. And, of course, scientists can now measure them better.

A special cause for concern is particulate matter less than 10 micrometres ($10^{-5}$ m) in diameter, known as $PM_{10}$. Such particles stay longer in the atmosphere than heavier particles, being mostly removed by precipitation rather than gravity. More important, while larger particles are generally filtered out by the small hairs in the nose and throat, $PM_{10}$ can settle in the bronchi and lungs, $PM_{2.5}$ can penetrate directly into the lungs, and $PM_1$ or smaller – $PM_{0.1}$ is the typical soot particulate matter emitted by modern diesel engines – can penetrate into the alveolar region of the lungs. Moreover, diesel soot particles carry carcinogenic chemicals adsorbed on their surface.

Grass and tree pollen – the cause of hay fever during certain seasons – is much larger, with a grain size between 15 and 100 micrometres ($PM_{15-100}$). Pollen does not enter the lungs and quickly settles under gravity: in a closed room most grains settle within 25 minutes. The problem pollen comes from species that rely on the wind, rather than insects, for reproduction. (Thus the pollen that sticks to bees is not the cause of hay fever.) These species must vastly overproduce pollen, for the bulk goes to waste. The giant ragweed releases 8,000 million pollen grains in just 5 hours.

Pollen concentration is measured by sucking ambient air and its particulates through a slit and passing it over an adhesive tape rotating at a set rate on a drum, then counting the trapped grains by microscope. The number is converted to pollen grains per cubic metre of air and publicly announced as low, moderate, high or very high. In the United Kingdom, June pollen counts in the Midlands, where the winds bring pollen from all directions, often exceed the 'very high' level (150+) by many times, reaching 1,000 and more.

| Level | Number of pollen grains (per cubic metre of air) |
|---|---|
| Low | 30 |
| Moderate | 30–49 |
| High | 50–149 |
| Very high | 150 or more |

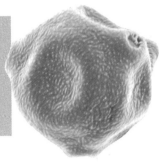

Above: **Birch pollen grain, magnified about 800 times.**

**Sir Humphry Davy (1778–1829), the chemist who invented the miners' safety lamp known as the Davy lamp in 1815. In underground collieries, explosions in pockets of methane (known as 'firedamp') triggered by conventional miners' lamps were wreaking havoc. Davy was aware that a gas 'will not explode in a small tube the diameter of which is less than the 8th of an inch.' His design (left) therefore replaced the glass chimney with a cylinder of brass gauze having holes small enough to prevent any explosion.**

# Sun Protection Factor

Sun worship seems to have existed since time immemorial. The Egyptian pharaohs claimed to be sons of the sun god Ra. King Louis XIV of France styled himself the Sun King in the 17th century. Plato likened the highest form of understanding, that of unchanging truths, to the Sun. And the poet Rabindranath Tagore (whose name Rabi means 'sun' in Bengali), wrote of the rising sun: 'Let its light reveal us to each other/ who walk on the same path/ of pilgrimage.'

Medically speaking, our bodies need the Sun, for without some exposure to its ultraviolet light, we cannot synthesize vitamin D, which is essential for calcium metabolism and the formation of new bone. A century ago, vitamin deficiency due to lack of light in the industrial cities of the West, caused widespread rickets, a disease characterized by bowleggedness. Today, doctors recommend about 15 minutes of direct exposure to the sun in the morning or evening for optimum vitamin D production.

Longer exposure to ultraviolet soon causes sunburn, and prolonged sunbathing – the contemporary form of sun worship – may lead to sunstroke and skin cancer. Fair-skinned people can to some extent protect themselves by developing the dark pigment melanin in their skin, in the form of a suntan, but unless they use artificial skin protection, the inevitable result of staying outdoors for an hour or two in fierce sunlight will be painful and ultimately skin-damaging sunburn.

The first known sunscreen lotion was olive oil, used by the ancient Greeks, and it was not very effective. Not until the 20th century did any reliable protection develop. In 1938, a Swiss chemistry student who was severely burnt while climbing in the Alps synthesized a 'glacier cream'. Samples still exist and tests show them to have a sun protection factor of 2, in other words a person wearing the cream can spend twice as long in the Sun before burning as without it. During the Second World War, when many soldiers were sunburnt, a thick petroleum-based product, 'red vet pet', similar to Vaseline was produced by a pharmacist. He tried it on his bald head, with limited results. Modern sunscreens generally work through two active ingredients: a chemical block consisting of an organic compound (such as benzophenone) that absorbs UV, and a physical block (such as titanium dioxide or zinc oxide) that reflects it – available either separately or, better still, in combination.

The sun protection factor (SPF) commercially available ranges as high as 30, which means that a protected person who would normally burn after 12 minutes could theoretically stay in direct sun for $30 \times 12$ minutes, or 6 hours. Claims for creams with higher factors, giving 'all-day protection', are unrealistic. Moreover, SPF is only an approximate guide: the actual level of protection will depend on the user's skin type, how much and how often the sunscreen is re-applied, whether the user's activity (for example swimming, as opposed to sunbathing) tends to remove the sunscreen, and what proportion of it has been absorbed by the skin.

Sun worship. The ancient Greeks tried olive oil on their skin as a protection from the Sun, but only in the 20th century was an effective sunscreen lotion chemically synthesized. A dark skin has some natural protection from ultraviolet solar radiation provided by its pigment melanin (from the Greek *melas*, meaning 'black'). But no one should expect to be able to stay out in strong sunlight all day, however dark their skin and however high the SPF number of their sunscreen lotion.

# Medical Prescriptions

A formal study of doctors' handwritten prescriptions found that five per cent were illegible – perhaps a lower figure than patients might have expected. But then prescriptions are intended not so much for patients as for pharmacists, or in earlier times apothecaries, who dispense the medicine and usually retain the paper prescription as a legal record. Pharmacists often get to know the prescribing styles of local doctors and can decipher their instructions. Of course, in theory it should be possible to do away with handwritten prescriptions if doctors are able to transfer prescriptions electronically to pharmacists using smartcards and the internet.

Whether paper or electronic, prescriptions will always require a specialized language, as unambiguous as possible and with standard abbreviations, for doctors to communicate to pharmacists the diverse treatment permutations of drug quantity, frequency, substrate (capsule, ointment, spray, etc.), circumstance (with or without food, for example) and other factors. The language used in scientific medicine is based on medieval Latin with many modern additions and variants for different countries, health authorities and hospitals. The table below includes only some of the commoner terms and abbreviations.

Left: **An apothecary making up a prescription. This early 19th-century caricature by A. Park 'conveys the ambiguity of public feelings towards the apothecary' (Roy Porter in *The Greatest Benefit to Mankind*). Note the skull and crossbones. Pharmacists are the modern apothecaries.**

| Abbreviation | Latin | Meaning |
|---|---|---|
| a.c. | ante cibum | before meals |
| ad lib. | ad libitum | as much as one wishes; to any extent |
| alt. h. | alternis horis | every other hour |
| b.i.d. | bis in die | twice daily |
| cap., caps. | capsula | capsule |
| dieb. alt. | diebus alternis | every other day |
| ex aq. | ex aqua | in water |
| gr | | grain |
| gtt(s) | gutta(e) | drop(s) |
| h.s. | hora somni | at bedtime |
| IV | | intravenous |
| M. | misce | mix |
| m. et n. | mane et nocte | morning and night |
| N.M.T. | | not more than |
| noct. | nocte | at night |
| o.d. | oculus dexter | right eye |
| omn. noct. | omni nocte | every night |
| p.c. | post cibum | after meals |
| p.r.n. | pro re nata | when required |
| pulv. | pulvis | powder |
| q | quaque | every |
| q. 1h | quaque 1 hora | every 1 hour |
| q.d. | quaque die | every day |
| q.i.d. | quater in die | four times a day |
| s.a. | secundum artem | according to art; use your judgement |
| s.o.s. | si opus sit | if there is a need |
| syr. | syrupus | syrup |
| tab. | tabella | tablet |
| troche | trochiscus | lozenge |
| ung. | unguentum | ointment |
| Y.O. | | years old |

# Disease Incubation

'Miasma' is defined in the *Oxford English Dictionary* as '*archaic*: an infectious or noxious vapour'. Until bacteria began to be accepted as the cause of diseases following the work of Louis Pasteur in the 1860s, it was widely believed that miasmas – foul stenches – were infectious. Astonishingly, Florence Nightingale still held to this outmoded view at her death in 1910. Contagion, she wrote, 'presupposes the existence of certain germs like the sporules of fungi, which can be bottled up and conveyed any distance attached to clothing, to merchandise… There is no end to the absurdities connected with this doctrine.'

Even so, the practice of quarantine (from the medieval French meaning a period of 40 days), began in Europe as early as the second half of the 14th century, after the Black Death. Forty days, it was assumed, would be sufficient for the symptoms of disease to appear on a ship or in a stranger. We now know that the incubation period of most diseases, though not all, is considerably shorter.

| Approximate Incubation Periods of Diseases | |
| --- | --- |
| Cholera | 1–3 days |
| Influenza | 1–4 days |
| Scarlet fever | 1–4 days |
| Common cold | 2–5 days |
| Severe acute respiratory syndrome (Sars) | up to 10 days |
| Poliomyelitis | 7–14 days |
| Measles | 9–12 days |
| Smallpox | 7–17 days |
| Generalized tetanus | 7–21 days |
| Chicken pox | 14–16 days |
| Mumps | 14–18 days |
| Rubella (German measles) | 14–21 days |
| Rabies | 2–6 weeks |
| Syphilis | 2–70 days |
| Variant Creutzfeldt-Jakob disease (vCJD) | up to 50 years (est.) |

Above and below left: **John Snow (1813–58) and his 1854 map of cholera deaths in Soho, central London, which launched the science of epidemiology. Snow, an up-and-coming anaesthetist, conducted extremely detailed investigations of the deaths, plotted the victims' addresses on a map of Soho and discovered a pattern: fatalities clustered around the Broad Street water pump. At Snow's request, the pump's handle was removed, and the epidemic subsided. In truth, the epidemic was already subsiding, but Snow's action saved further lives. Much more important in the long run was his conviction that the disease was waterborne, not miasmatic. This was proven right, when the pump's well was investigated and found to be contaminated by a nearby cesspit into which a cholera victim's nappies had been rinsed.**

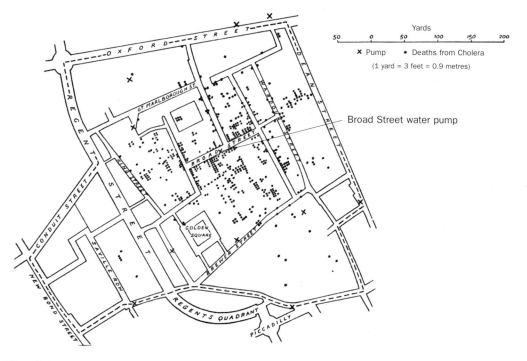

Yards

| 50 | 0 | 50 | 100 | 150 | 200 |

✕ Pump • Deaths from Cholera
(1 yard = 3 feet = 0.9 metres)

Broad Street water pump

# Pain

In the history of pain, a startling fact is that a generation or two of surgeons ignored anaesthesia after Humphry Davy, the discoverer of nitrous oxide in 1800, suggested its possible use in dentistry and surgery. 'When chloroform was introduced [in the 1840s] it was bitterly criticized as immoral – because it relieved pain', notes *The Oxford Companion to the Body*. 'Pain was not regarded as a physical malfunction but as part of the Universe.'

The first medicine labelled as a 'painkiller' was marketed in 1853. It is hard indeed to imagine that pre-analgesic world, notwithstanding the pain in our modern world. As a child in the late 1960s, I sometimes had teeth drilled by a dentist without an injection; now I always accept the proffered jab. Has my body really become more sensitive to pain as an adult, or is it that scientific progress has led me to expect that life should be pain-free?

Pain is certainly the most common symptom seen by general practitioners. Yet it is very difficult to measure, because it is so subjective. For a long time, doctors wrongly thought of pain only in terms of self-protective mechanisms resulting from injury – the agony of a banged elbow forcibly reminding you not to bang it again. But we know that injury may often occur without pain, and pain without injury. As the physiologists Ronald Melzack and Patrick Wall write in *The Challenge of Pain*, 'Why are pain and injury not always related?'

Using the questionnaire (above), adapted for a study of narcotic drugs, patients were

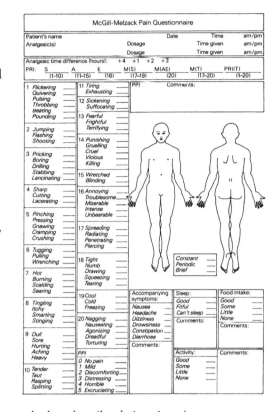

McGill-Melzack Pain Questionnaire. Although pain is intensely subjective, it can be objectively measured to some degree by carefully questioning sufferers. (See text for fuller explanation.)

asked to describe their pain using twenty different sets of words. Each word was given a rank value within the set, and the sum of the rank values was the 'pain rating index' (PRI), which was supplemented by the 'present pain intensity' (PPI) based on a scale of 0 ('no pain') to 5 ('excruciating'). When the questionnaire was administered to those with one of eight syndromes ranging from non-terminal cancer to toothache, statistical analysis showed that 'each type of pain has unique qualities which are described by a distinctive set of words' (Melzack and Wall), such as 'throbbing' for toothache and 'cramping' for labour pain. The PRIs ranged from 41 (amputation of a digit), through 26 (non-terminal cancer) to 16 (a simple sprain).

# Stress Factors

Outside physics and engineering, and perhaps plant science, there is no widely agreed way to measure stress. All attempts to quantify human stress therefore have to be treated with caution and scepticism.

To begin with, stress is not like pain: no one welcomes pain, except masochists. But an activity or situation that is stressful for one individual can be enjoyable and relaxing for another – think of exams, public speaking and sports like rock climbing. Challenges make life meaningful to some people, but make others want to run away. By way of a mundane example, I have commuted in London for years by bicycle, and like it for reasons of convenience, exercise and cost, despite my occasional outbursts of anger at the behaviour of motor vehicles, other cyclists and pedestrians. But I know many people who would find regular big-city cycling extremely stressful.

Secondly, who is to be the arbiter of stress – the person who reports being stressed or some 'impartial' professional such as a doctor or psychologist? If an employee reports a feeling of stress, and the employer can find no reasonable cause for it, who is correct?

Thirdly, are scientific measures of stress necessarily more valid than, say, opinion and attitude surveys? Physiological indicators like pulse rate, glandular secretions and brain activity can be measured easily enough with instruments, but they may not have any obvious connection with mental stress. Certainly, polygraphs (lie detectors) have a very poor record of telling truthful people and liars apart (see pp. 201–202).

Anyway, how can one validly compare and quantify very different kinds of stress, such as having a major illness, having a baby and moving house, not to mention disasters like earthquakes and wars? And what about simultaneous stresses, where, say, worry about a difficult child provokes rows with a spouse? The permutations in the causes of stress preclude scientific measurement of it.

| | |
|---|---|
| Death of a spouse | 100 |
| Divorce | 73 |
| Marital separation | 65 |
| Jail sentence | 63 |
| Death of a close family member | 63 |
| Personal injury or illness | 53 |
| Marriage, engagement or living together | 50 |
| Loss of a job | 47 |
| Marital reconciliation | 45 |
| Retirement | 45 |
| Change in health of a family member | 44 |
| Pregnancy | 44 |
| Sex difficulties | 39 |
| Birth of a baby | 39 |
| Changes in financial position | 38 |
| Death of a close friend | 37 |
| Change to a different job | 36 |
| Taking out a large mortgage or loan | 31 |
| Promotion or demotion at work | 29 |
| Son or daughter leaving home | 29 |
| Trouble with in-laws | 29 |
| Outstanding personal achievement | 28 |
| Spouse begins or stops work | 26 |
| Beginning or end of school or college | 26 |
| Trouble with the boss | 23 |
| Change in work hours or conditions | 20 |
| Moving house | 20 |
| Change in school or college | 20 |
| Change in recreation | 19 |
| Change in social activities | 18 |
| Change in sleeping habits | 16 |
| Change in eating habits, e.g. dieting | 15 |
| Holiday | 13 |
| Christmas | 12 |
| Minor violations of the law | 11 |

Stress factors. Although stress is very hard to define, psychologists, health professionals, employers, magazine and newspaper editors, and indeed all of us, try to measure it. The subject is of such importance and interest that efforts will always be made to rank stress in a meaningful way. This particular 0–100 ranking is merely the assessment of one professional and should not be taken too seriously, neither does it claim to be complete. How could it be, given the all-embracing nature of stress?

# Textiles and Togs

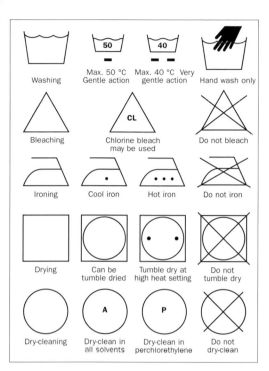

Washing | Max. 50 °C Gentle action | Max. 40 °C Very gentle action | Hand wash only

Bleaching | Chlorine bleach may be used | Do not bleach

Ironing | Cool iron | Hot iron | Do not iron

Drying | Can be tumble dried | Tumble dry at high heat setting | Do not tumble dry

Dry-cleaning | Dry-clean in all solvents | Dry-clean in perchlorethylene | Do not dry-clean

Labels on clothes, besides giving the size (in the imperial, metric or other system, depending on the country of purchase) and the content of the fabric, also give cleaning instructions. These are sometimes in words but mainly in the symbols of the International Textile Care Labelling Code. Its five basic symbols, as shown at the left of this diagram, specify the five operations of washing (wash tub), bleaching (triangle), ironing (iron), drying (square) and dry-cleaning (circle). To these may be added more specific instructions: numbers in degrees Celsius for washing temperature and horizontal bars for washing action; letters for chlorine bleach and different dry-cleaning chemicals; and dots for ironing and drying temperature. A cross through the symbol indicates that the operation is prohibited. Less common square

symbols (not shown) stand for 'drip dry', 'hang-line dry' and 'dry flat'.

A specification found on most duvet labels is the 'tog value'. This unit of thermal resistance generally ranges from 4.5 to 15 tog. The higher it is, the higher the thermal resistance and the warmer the duvet. Tog value is roughly proportional to a duvet's thickness, or rather to the amount of air trapped in the duvet, independent of the type of filling. (Still air, an excellent insulator, generally provides 99 per cent of the volume of filling.) A tog of 4.5 indicates light summer use, a tog of 15 – 'superwarm plus' – is otiose for houses with central heating.

'Tog' came from the common English slang word for clothes. The unit was invented in the 1940s for the convenience of textile workers at the Shirley Institute in Manchester, and publicly launched in the 1960s. Scientifically, thermal resistance is defined as the temperature drop across an insulator divided by the heat flow through it, and therefore its SI unit is degrees kelvin divided by watts per square metre – with a value of 0.1 for men's suiting. The tog value is defined as 10 times the SI value – in this case 1. The tog scale is linear, not logarithmic, hence a duvet of tog 15 has 15 times the thermal resistance of a blanket of tog 1. (The tog values of blankets are too similar to be usefully stated on a label.)

A duvet's tog value is measured with a togmeter. A sample is placed on an electrical hot-plate apparatus, which measures the temperature drop with a thermocouple and the heat flow from the heater's power input.

| Clothing | Tog value |
|---|---|
| shirting | 0.7 |
| underwear 'thermal' | 0.2–0.4 |
| underwear | 0.4–0.8 |
| suitings | 1.0 |
| sweaters | 1.0 |
| blankets | 1.0–2.0 |
| duvets | 4.5–15 |

Above: **Tog values of clothing.**

Left and above: **The International Textile Care Labelling Code.**

**Temperance** by Pieter Bruegel the Elder, 1560. A new obsession with measurement takes hold of society. One person measures a pillar with a divider and plumb-line, another a work of art with a right angle, yet another angular separation by sighting along a stick with a wheel at its end. In the centre, Temperance has a spring-driven clock on her head and the blade of a windmill under her foot, a pair of spectacles in one hand and a pair of reins in the other connected to a bit in her mouth, while her girdle is made of a knotted snake. These suggest a sense of purpose and self-control arising from society's new respect for measurement.

# Calendars

All working calendars, whether they were the Babylonian, Julian or Mayan calendars of the ancient world, or the Gregorian, Jewish or Muslim calendars of our own time, were based on either the movements of the Earth versus the Sun or the Moon around the Earth, or on both movements. They monitored the turning of the seasons, the phases of the Moon and the rotation of the Earth, which determined the year, the months and the day, respectively. These calendars therefore had to contend with three inconvenient facts: the lunar month is just under 29.5 days long – not 30 days; the solar year is just under 365.25 days long – not 360 days; and the solar year is 12.4 lunar months long – not 12 months. The motions of the Earth and the Moon do not provide whole-number ratios between our perceived time periods of day, month and year.

Leap years – years with an extra day – were required to keep the calendars in step with solar years. In the Julian ('Old Style') calendar, dating back to Julius Caesar in 46 BC, the fourth year was always a leap year. The mean, or average, year was one quarter of the sum of (365 + 365 + 365 + 366) days, which equals 365.25 days. But since the solar year is in fact 365.2422 days, the Julian calendar gradually accumulated too many days and fell behind the solar year.

In 1582, Pope Gregory XIII introduced a calendar reform, the 'New Style' calendar. He suppressed 10 days; in Catholic Europe, the day after 4 October 1582 was 15 October. Then, to stop a future lag behind the solar year, a few leap days had to be suppressed periodically. Gregory provided that, of the centenary years (1600, 1700, and so on), only those exactly divisible by 400 should be counted as leap years. Thus 1600 was a leap year, 1700, 1800 and 1900 were not, whereas 2000 was a leap year; this removed 3 leap days every 400 years. Taken over 2000 calendar years, it meant the removal of 15 leap days from the normal 500 leap days (based on one every 4 years), which gives 485 leap days. Add 485 to the 730,000 (2000 × 365) regular days in 2000 years, and you get a total of 730,485 days in 2000 years, which is 365.2425 days per year – a mere 0.0003 days more than the solar year, and a negligible error.

In anti-Popish England and Wales (and colonial America), because of a delay until 1752 in adopting the Gregorian calendar, 11 days had to be suppressed; 2 September 1752 was followed by 14 September. Dates before 1752 must take account of the change.

A French Revolutionary *calendrier perpétuel*. In fact, it lasted a mere thirteen years. It was introduced in October 1793 and began with the year II, and it was abolished in the year XIV by Napoleon, who decreed that 1806 would start on 1 January. Not only had the new calendar been universally unpopular in France with its ten-day working week, Napoleon wanted papal legitimacy for his regime and the Catholic church wanted its Sundays and saints' days back. The months of the calendar were named by the poet-dramatist Fabre d'Eglantine (who was guillotined in 1794). There were 12 months of 30 days each. The missing 5 days of the conventional 365-day calendar year were designated a festival.

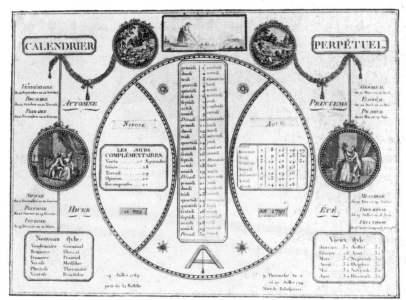

September hath xix Days this Year. 1752

First Quarter, *Saturday* the 15th, at 1 aftern.
Full Moon, *Saturday* the 23d, at 1 aftern.
Last Quarter, *Saturday* the 30th, at 2 aftern.

| 1 | f | Giles Abbot | 5 38 | 6 22 | secret | □ ♃ ☿ | 5 |
| 2 | g | London Burrt | 5 40 | 6 20 | memb. | Wind, | 6 |

According to an Act of Parliament passed in the 24th Year of his Majesty's Reign, and in the Year of our Lord 1751, the Old Style ceases here, and the New takes place; and consequently the next Day, which in the Old Account would have been the 3d, is now to be called the 14th; so that all the intermediate nominal Days from the 2d to the 14th are omitted, or rather annihilated this Year; and the Month contains no more than 19 Days, as the Title at the Head expresses.

| 14 | e | Holy Cross | 5 42 | 6 2? | thighs | and stor- | 7 |
| 15 | f | Day decreas'd | 45 | 2c | hips | my Wea- | 8 |
| 16 | g | 4 hours | 46 | 18 | knees | ther. | ☽ |
| 17 | A | 15 S. aft. Tri. | 48 | 1? | and | Fair and | 10 |
| 18 | b | Day br. 3. 45 | 50 | 14 | hams | seasonab. | 11 |
| 19 | c | Clo. flow 6 m. | 52 | .12 | 'egs | ♂ ☉ ☍ | 12 |
| 20 | d | Ember Week | 54 | 1c | ancles | ♂ ♀ ☿ | 13 |
| 21 | e | St. Matthew, | 56 | 8 | feet | Rain and | 14 |
| 22 | f | | 56 | 6 | toes | Windy. | 15 ● |
| 23 | g | Eq. D. & N. | 5 58 | 4 | head | ♂ ☉ ☍ | |
| 14 | A | 16 S. aft. Tri | 6 0 | 2 | and | | 17 |
| 25 | b | Day dec. 4, 34 | 2 6 | c | face | □ ♃ ♀ | 18 |
| 26 | c | S. Cyprian | 4 5 | 5? | neck | ♂ ☉ ☿ | 19 |
| 27 | d | Holy Rood | 6 | 5? | .hroat. | Inclin. to | 20 |
| 28 | e | Clo. flow 9 m. | 8 | 5? | arms | ♂ ♂ ☿ | 21 |
| 29 | f | St. Michael | 10 | 5c | should. | wet, with | 22 |
| 30 | g | St. Jerom | 12 | 4? | breast | Thunder. | ☾ |

For example, the birth year of Isaac Newton is usually given as 1642, but is sometimes given as 1643, because Newton was born on 25 December 1642 (Old Style), which is 4 January 1643 (New Style). George Washington was born on 11 February 1732 (O.S.) or 22 February 1732 (N.S.).

It is widely believed that English people rioted against the Popish imposition and 'loss' of 11 days from their lives in the 1752 calendar change. But Robert Poole, a historian who has searched for reliable evidence of these 'calendar riots', has found nothing solid, though there was a certain amount of discontent about the upset to the rhythm of familiar festivals, notably Christmas. 'Most of the people at this period measured time not by calendar dates but by the passage of festivals and anniversaries; diary use was still largely confined to the gentry, tradespeople and professional classes.' The first time many would have become aware of the calendar change was when Christmas was proclaimed in church on what had been 14 December. The only explicit reference to the notorious 'lost' 11 days appears to be William Hogarth's satirical painting of 1755, *An Election Entertainment*, which shows a wounded Whig bludgeon-man with a placard grabbed from a Tory that reads 'Give us our Eleven Days'. However the painting refers not to any calendar riots but to a hard-fought election in Oxfordshire in 1754, in which the calendar reform may have played a minor part. By contrast, the almanacs of 1752 carefully explained what had happened to the 'missing' 11 days.

Above: *An Election Entertainment* by William Hogarth, 1755 (detail). The placard on the floor under the wounded man's foot reads 'Give us our Eleven Days' – a reference to calendar reform.

Left: A British almanac for 1752 explains the change from the Julian (Old Style) to the Gregorian (New Style) calendar. September has only 19 (XIX) days.

# Time Zones

The establishment of Greenwich Mean Time (GMT) in 1884 created official international time zones and the international date line through the Pacific Ocean 180 degrees from the prime meridian at Greenwich (seen in its present form at right). Eastbound travellers crossing the line must set their watches back by a whole day and westbound travellers must advance theirs by a day – a fact that saves eastbound Phileas Fogg in Jules Verne's *Around the World in Eighty Days* from losing his famous bet. The boundaries of time zones and the date line are irregular (see map), to take account of national decisions. Some large nations, for example India and China, impose a single time throughout the country, but others do not: Russia has eleven time zones.

Summer Time, for daylight saving, was first suggested by a London builder, William Willett, in the early 20th century; his idea was that people should put their clocks forwards by one hour to catch the early morning light in the spring and summer. Wartime economy ensured its adoption in 1916, first in Germany and Austria-Hungary, then in Britain, where British Summer Time (BST) has been in force since 1922. Following European Union agreement, Summer Time starts at 1 a.m. GMT on the last Sunday in March and finishes at the same hour on the last Sunday in October. Most other nations have their own form of daylight saving, although US states may exempt themselves – as has happened in Indiana, which straddles a line between time zones, in Arizona and in Hawaii.

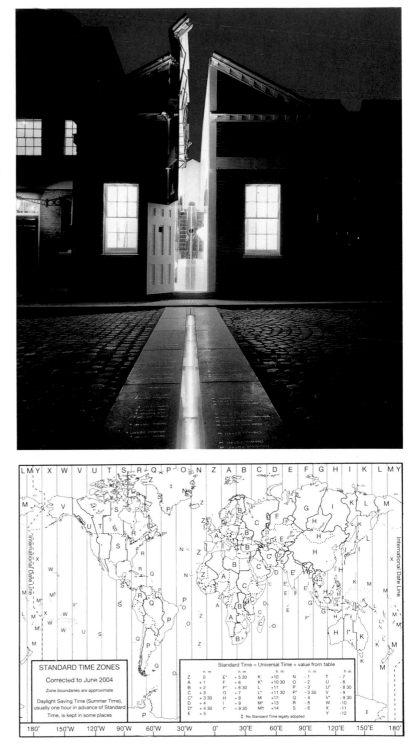

**STANDARD TIME ZONES**

Corrected to June 2004

Zone boundaries are approximate

Daylight Saving Time (Summer Time), usually one hour in advance of Standard Time, is kept in some places

Standard Time = Universal Time + value from table

| | h m | | h m | | h m | | h m |
|---|---|---|---|---|---|---|---|
| Z | 0 | E* | + 5 30 | K | +10 | N | - 1 | T | - 7 |
| A | + 1 | F | + 6 | K* | +10 30 | O | - 2 | U | - 8 |
| B | + 2 | F* | + 6 30 | L | +11 | P | - 3 | U* | - 8 30 |
| C | + 3 | G | + 7 | L* | +11 30 | P* | - 3 30 | V | - 9 |
| C* | + 3 30 | H | + 8 | M | +12 | Q | - 4 | V* | - 9 30 |
| D | + 4 | I | + 9 | M* | +13 | R | - 5 | W | -10 |
| D* | + 4 30 | I* | + 9 30 | M† | +14 | S | - 6 | X | -11 |
| E | + 5 | | | | | | | Y | -12 |

‡ No Standard Time legally adopted

# Postal Codes

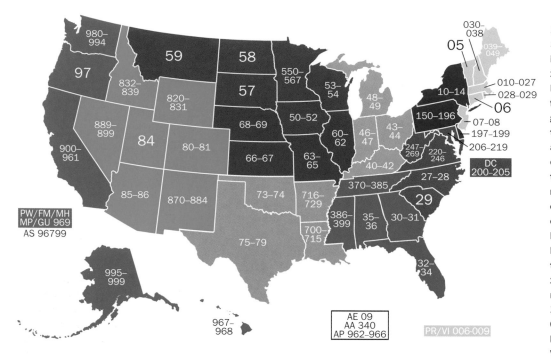

Postal districts and codes might seem to be neutral subjects. But they acquire economic, social and cultural significance. In the United States, the Zip codes 90210 (Beverly Hills, California) and 02138 (Cambridge, Massachusetts) enjoy a cachet deriving from Hollywood films and Harvard University. *National Geographic* magazine even had a regular monthly feature about the community with a particular Zip code.

Postal codes were briefly introduced in Germany in 1941, but first became established in the United Kingdom. London had acquired numbered postal districts in 1857–8, following Rowland Hill's division of the city by the compass into ten districts: EC (Eastern Central), WC, NW, N, NE, E, SE, S, SW and W. In 1866, after a report by the surveyor Anthony Trollope (the novelist), NE was merged with E district, and then S was divided between SE and SW districts. In 1917, during the war, to assist women sorters standing in for men, a serial number was added to identify London boroughs, for example N1 for Islington, SW6 for Fulham. Finally, from 1959 up to the early 1970s, with mechanization of mail sorting, the current postcodes were introduced. The principle can be illustrated by the postcode PO1 3AX. PO1 is the outward code indicating Area PO (of which there are 121 in the UK) and District 1 (about 20 postcode districts in an area); 3AX is the inward code specifying Sector 3 (with about 300 addresses in a sector) and within it Unit AX (about 15 addresses per unit). Though there are many exceptions, including the postcodes for London, in general a UK postcode's first letter or two corresponds to the location it describes, such as PO for Portsmouth.

**Zip code zones.** The 5-figure Zip code was introduced in the USA in 1963 as part of the Zone Improvement Plan. The first digit, 0-9, represents a group of states (see map); the second and third digits a region within the group or maybe a large city; and the fourth and fifth digits more specific areas, for example small towns or city regions. Thus, within New York City, Staten Island and The Bronx have the Zip codes 10300-10399 and 10400-10499 respectively. The lowest Zip codes, beginning with 0, are in New England, Puerto Rico and the US Virgin Islands, and the highest in California, Hawaii and Alaska. US government agencies in and around Washington DC have codes beginning with 20200 and 20599, regardless of their exact location. The system has many flaws, and in 1983, an expanded 9-figure Zip code had to be introduced, known as 'Zip+4' or 'add-on codes'. (Continental Europe, by contrast, tends to use a 4- or 5-digit system, which indicates region and sub-region, sometimes prefixed by a letter standing for the country, such as D for Deutschland/Germany.)

# Opinion Polls

Opinion polls, market research and social surveys are ubiquitous tools of politicians, commercial organizations and academics. Indeed, 'The study of voting behaviour entered political science from market research by way of electoral polling', writes the historian Theodore Porter in *Trust in Numbers*. However they suffer from several weaknesses that leave their conclusions open to challenge.

One difficulty is that data collection is expensive, especially internationally, and so sample size is sometimes insufficient. Secondly, responses are difficult to standardize, because people gave different answers to logically equivalent questions merely as a result of different phrasing or different ordering of the questions. Thirdly, respondents may exaggerate or lie. In a recent survey of Americans conducted by the National Opinion Research Center at the University of Chicago, 24–30 per cent claimed to go to church at least weekly, but the researchers noted that as a group the respondents typically overstated their attendance by up to 70 per cent. The more sensitive the issue, the more likely respondents are to dissemble.

Academics – in contrast to professional pollsters with no personal stake in their research – try to take account of the second and third of these problems by actively encouraging people to respond in their own words, rather than adopting a rigid 'tick-in-the-box' methodology. But this deeper approach, besides requiring more time from both questioner and respondent,

leaves the responses hard to compare. Even so, academic attitude surveys are undeniably fascinating – consider the map of world happiness below.

The online world may have made polling cheaper for pollsters, but probably no more reliable. Certainly it has not resolved the debate over electronic voting that took off after the fiasco of the 'hanging chads' in many Florida voters' papers in the US presidential election of 2000. 'To have a fair, democratic election, there has to be a visible, transparent way of performing recounts and confirming that ballots have been cast correctly', says Rebecca Mercuri, an academic authority on electronic voting systems. She, and many others, believe the only way to ensure fairness is for the electronic voting machine to print out a paper copy at the time of voting – which the voter may review before casting the vote, and which officials may recount in case of a dispute between closely matched candidates.

Below: **A map of world happiness, 2006. It is the result of analysing the findings of over 100 studies worldwide, which questioned 80,000 people, and incorporates data from Unesco, the CIA, the New Economics Foundation, the WHO, the Veenhoven Database, the Latinbarometer, the Afrobarometer and the UNHDR. The compiler, social psychologist Adrian White, notes a strong positive correlation between happiness and health, closely followed by wealth and education. Scandinavia tops the list for happiness, with Denmark at no. 1. The USA comes 23rd, ahead of Germany, the UK and France. The Asian countries, China (82nd), Japan (90th) and India (125th) come low, perhaps surprisingly.**

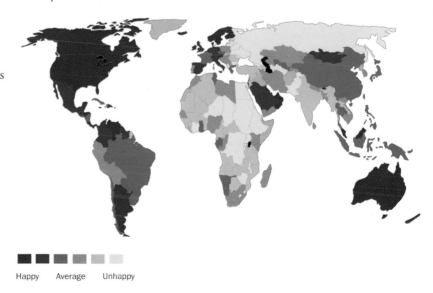

Happy    Average    Unhappy

# Censuses

Censuses are inherently political and often controversial. For example, King David ordered that 'Israel and Judah should be counted' by army officers. After nine months and twenty days, they reported 800,000 'able-bodied men capable of bearing arms' in Israel and 500,000 in Judah. Then David decided his census was sinful and asked God for punishment. He was offered a choice: famine, slaughter by enemies or pestilence, each lasting three months. He chose pestilence, and 70,000 people perished. Even in the 18th century, this biblical precedent was cited in Parliament against plans to enumerate the English population.

Pestilence was the reason for the first truly statistical analysis. So-called 'bills of mortality', reporting deaths over a given period, were compiled in 16th-century England during plagues. After the epidemic of 1603, the bills came to be published regularly, even when there was no plague. In 1662, a merchant John Graunt made the first estimate of London's population – 384,000 – by analysing the bills on certain assumptions: that the annual number of christenings approximated the birth rate; that women between the ages of 16 and 40 gave birth bi-annually; and that the average family unit consisted of 8 people: mother, father, 3 children and 3 servants or lodgers.

Today, the Indian government uses a much-disputed 'pugmark census' for counting tigers: forest officers roam the national parks for a week or two looking for paw prints. Then in 2005 it became clear that there were no tigers left in one particular reserve, where the official figure was 16–18 animals. The storm of protest forced the government to use sampling methods based on remotely triggered cameras in addition to counting pugmarks.

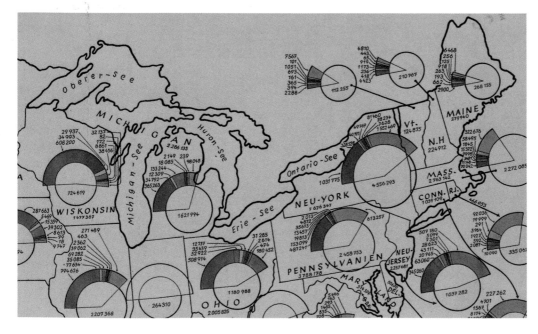

Nazi 'census' of the United States, 1940. This map shows part of a secret analysis of the US population by the German government, based on the official US census of 1930. Each circle breaks down a state's population of white immigrants by their country of descent; those of German, Austrian, Dutch and Belgian descent are all included in the red portion. Nazi funds were then channelled to these populations, in the hope of using their potential loyalty to bring pressure on the US government to stay out of the war in Europe. Wisconsin was clearly a more promising target for Nazi propaganda than Massachusetts. 'The campaign had a considerable impact on restraining President Roosevelt' (Peter Barber in *The Map Book*). The US did not join the European war until the end of 1941.

# ID, Truth and Lies

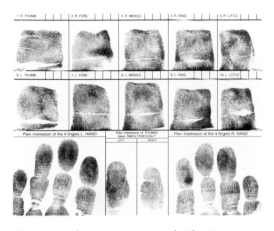

How to authenticate a person's identity beyond reasonable doubt has been an issue since the beginning of civilization. A seal can be stolen, a signature faked, a photograph doctored, a computer password hacked. What is needed is a biometric indicator: an irremovable, easily measurable body feature unique to an individual. For the past century or so, this role has been filled by fingerprints.

The distinctiveness of finger marks had been noticed as far back as 1684 in the journal of the Royal Society, and in the early 19th century a British engraver had 'signed' his books on birds with his fingerprints, but it was not until the last decade of the 19th century that fingerprinting caught on as a means of legal identification. It happened not in Europe but in colonial Bengal, as a consequence of 'the sheer magnitude of the colonial problem of identifying every individual who might buy a piece of land, draw a pension, or sign some kind of contract', writes Chandak Sengoopta in his history of early fingerprinting, *Imprint of the Raj*. An administrator, William Herschel, initiated such fingerprinting; a scientist in London, Francis Galton, classified the fingerprints, beginning with three main categories, arches, loops and whorls; and a police chief, Edward Henry, applied the technique, first in Bengal and then in London, by matching fingerprints from crime scenes with fingerprints taken from known criminals.

But how reliable, really, is fingerprinting? How certain can we be about a supposed match? Astonishingly, for a century no one bothered to establish the error rate. Then, as a result of a court case in 1999, the American FBI was compelled by a judge to investigate fingerprinting forensically. Its report claimed that the probability of a false match was effectively zero, but this claim was rejected by the director of the US National Biometric Test Center as guesswork. From various pieces of recent evidence, it is clear that while fingerprinting is an extremely useful method of identification, it is not as irrefutably convincing as law courts have generally chosen to accept.

In the near future, biometrics other than fingerprinting may become part of everyday life. DNA profiling is already with us (see pp. 177–8), and iris recognition may follow – for example, when withdrawing money from a bank's cash machine, logging on to a computer, entering a building or passing through immigration control at an airport. The iris of the human eye has unique

Sir Francis Galton (1822–1911), the scientist who first classified fingerprints in the late 19th century. His methods were first applied to provide crucial evidence of a murder suspect's identity in Bengal in 1897. A fingerprint found in the room of the murder matched a fingerprint in a police file (below). However, recent investigations of fingerprint evidence show it to be very occasionally unreliable.

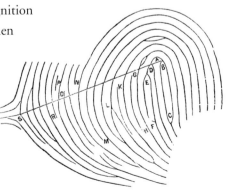

identifying patterns of pigmentation that cannot be faked or altered (unlike faces), and these can be scanned and digitized. A particular iris pattern can then be converted into a unique identifying numerical code that should enable a camera linked to a database to recognize a person.

However, experience with the polygraph, or lie detector, recommends caution in introducing new biometrics. Lie detectors have been around for decades, and are enthusiastically used by the US government, but there is no consensus among psychologists that they actually work. The original idea was to measure a person's breathing, blood pressure and electrodermal responses, that is the way in which sweat changes the skin's electrical conductivity in the palm of the hand, under questioning. Now, MRI brain scans have been introduced too, raising high expectations. Theoretically, all these measures are supposed to show increased activity when an innocent person

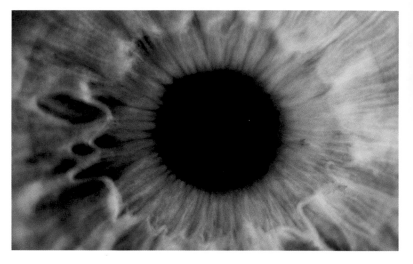

answers a control question (such as 'Have you ever told a lie to get out of trouble?') but not when an innocent person answers a question directly relevant to the charge; the latter question should, of course, disturb guilty persons and show increased activity in their bodies and brains.

Four studies of lie detectors published in scientific journals since 1976 do not support the theory. Innocent people are very nearly as likely as guilty people to be shown to be 'deceptive'. While there may possibly be some subtle changes in brain activity that correlate with deception, such data are significant only when pooled and not when applied to individuals. An American professor of psychology, David Lykken, regards polygraph tests as 'strongly biased against the truthful person' and warns that 'Marrying the myth of the lie detector with the mystique of the computer has spawned a progeny of mythlets that have compromised US intelligence findings and victimized honourable people.'

Above: **Human iris – the 21st-century fingerprint? But there are serious drawbacks to iris recognition. Significant numbers of people have droopy eyelids, squints, lazy eyes and large pupils that will interfere with computerized recognition; there is even a rare condition, aniridia, as a result of which about 1 in 75,000 people have no iris. And how will everyone update their iris image, which may change over time?**

# Race

When Einstein became world famous in 1919, he received an appeal for help from a group called the Central Association of German Citizens of the Jewish Faith. He replied: 'I am neither a German citizen, nor is there anything in me which can be designated as 'Jewish faith'. But I am a Jew and am glad to belong to the Jewish people, even if I do not consider them in any way God's elect.' The unbelieving Einstein liked to refer to Jews as his 'tribal companions'. Even he recognized the reality of tribe and race, while courageously attacking racism.

The study of race was tainted in its formative years. By measuring the physiognomy of races, anthropologists tried to define race-based correlations between biology, intelligence and culture. This led, via developments in genetics, to eugenics – the idea that races could be improved by sterilizing the unfit. Sterilization continued in Scandinavia and parts of Canada and the United States until the early 1970s.

Hence, when *Scientific American* tackled race in 2003, its editors asked gingerly: 'Does race exist?'. They quickly answered: 'If races are defined as genetically discrete groups, no. But researchers can use some genetic information to group individuals into clusters with medical relevance.'

Recent evidence about race comes from sequencing the human genome. Michael Bamshad, who studied the genetic makeup of 565 people born in sub-Saharan Africa, Asia and Europe, explains: 'To determine the degree of relatedness among groups, geneticists rely on tiny variations, or polymorphisms, in the DNA – specifically in the sequence of base pairs, the building blocks of DNA. Most of these polymorphisms do not occur within genes, the stretches of DNA that encode the information for making proteins... Accordingly, these common variations are neutral, in that they do not directly affect a particular trait. Some polymorphisms do occur in genes, however; these can contribute to individual variation in traits and to genetic diseases.'

A class of polymorphism known as *Alu* is passed unchanged from generation to generation for eons. If the same *Alu* sequence occurs at the same spot in the genomes of two different people, they must have a common ancestor and belong to a related group. Having removed all the identifying labels (place of origin and self-reported ethnicity) from their genetic samples, Bamshad and co-workers analysed only *Alu* polymorphisms, and discovered 4 different groups. They now restored the labels and found that 2 of the groups came entirely from sub-Saharan Africa (with 1 group made up almost exclusively of Mbuti Pygmies), and the other 2 groups comprised only individuals from Europe and East Asia, respectively. Some 60 *Alu* polymorphisms sufficed to match individual and continent of origin with 90 per cent accuracy; 100 were required for nearly 100 per cent accuracy.

Of course a polymorphism in common is just that. When a person's whole genome is considered, it is clear that it may vary as much between individuals of the same race as it does between members of different races.

Next page: **The Human Race Machine. Is race more than skin deep? The images on the next page were created by the American artist Nancy Burson from a photograph of a white woman, by adding and subtracting various racial features to show what she might look like as a member of another race. Sitting in front of the machine, individuals can map the features of their own faces, select an equivalent 'race' – and reflect on their new appearance. Burson herself states firmly: 'There is only one race, the human race... There is no gene for race.'**

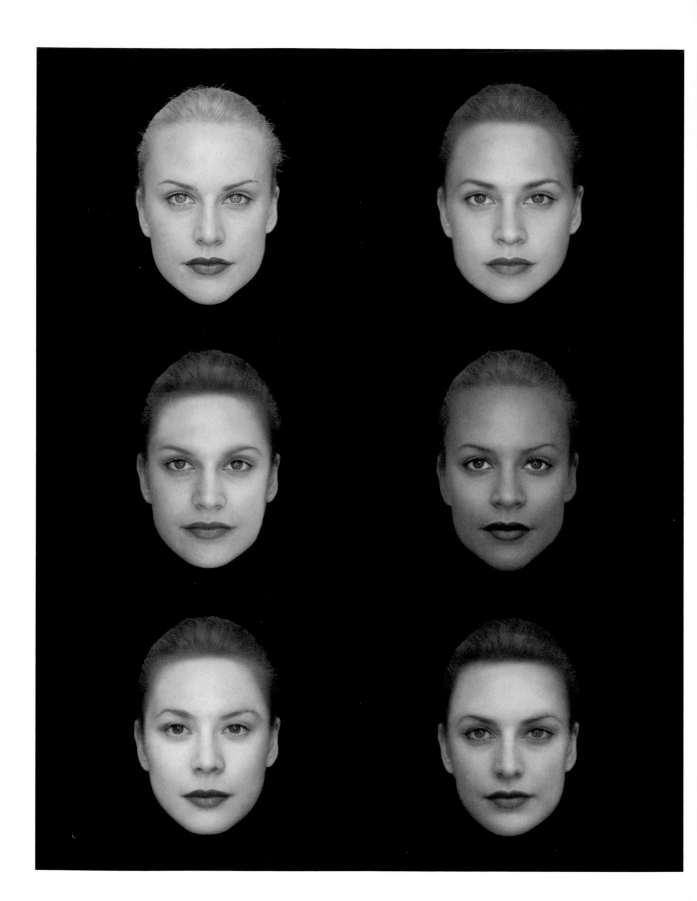

# Military Rank

The ancient Egyptian army had ranks in the 2nd millennium BC, as did the ancient Greeks, but it was ancient Rome that first formalized military rank. From the early 1st century BC, a legion – nominally 6,000 men – was commanded by a legate with 6 tribunes immediately beneath him. It was divided into 10 cohorts, each consisting of 6 centuries, which actually consisted of between 60 and 160 legionaries; the century was further broken down into tent messes (*contubernia*) of around 8 legionaries. A century was commanded by a centurion, assisted by junior officers.

Current English-language military ranks derive from Renaissance mercenary titles (for example, corporal, sergeant, captain and general), the era of the Napoleonic wars (colonel, lieutenant, major and marshal), and from the Second World War (specialist ranks, such as technical officer in radar).

Most armed services are divided into commissioned officers (general, colonel and so forth, who hold a government warrant conferring their rank), non-commissioned officers (sergeant, corporal, petty officer and others), and ranks with no authority to command (private, marine, airman, seaman and so on). And all armed forces make use of ranks in some form, except during periods of ideological experiment, as in the Soviet Red Army 1918–35, the Chinese People's Liberation Army 1965–88 and the Albanian Army 1966–91, each of which eventually restored rank after experiencing operational difficulties.

**Current British Army officer rank insignia. 1 Field Marshal, 2 General, 3 Lieutenant General, 4 Major General, 5 Brigadier, 6 Colonel, 7 Lieutenant Colonel, 8 Major, 9 Captain, 10 Lieutenant, 11 Second Lieutenant. (The rank of field marshal is now in abeyance.)**

## Senior Ranks in the Second World War

| Germany* | Japan | UK | USA | USSR |
|---|---|---|---|---|
| Field Marshal | Field Marshal (honorary rank) | Field Marshal | General of the Army | Marshal of the Soviet Union |
| Colonel General | General | General | General | Army General |
| General of the Infantry, Artillery, Engineers, etc | Lieutenant General | Lieutenant General | Lieutenant General | Colonel General |
| Lieutenant General | Major General | Major General | Major General | Lieutenant General |
| Major General | n/a | Brigadier | Brigadier General | Major General |

*Wehrmacht rank (the SS and Waffen-SS had separate ranks)

# Gun Calibre

The ring in Egyptian monumental hieroglyphic inscriptions that encircles and identifies royal names was named a cartouche by French soldiers of Napoleon's army, because to their eyes the ring looked rather like the shape of a gun cartridge – a *cartouche* in French. The diameter of a cartridge was known in French as its *calibre*, and this word too has been absorbed into English. Its earlier derivation appears to be from Italian *calibro* or Spanish *calibre* coming from the Arabic *qalib*, meaning 'mould', and ultimately from *kalapous*, the ancient Greek word for a shoemaker's last. This would make sense, given that gun barrels were once made from iron beaten into shape on a semi-circular anvil.

The calibre of large guns – artillery pieces – was historically stated not in terms of the external diameter of the projectile or the internal diameter of the barrel, but rather in terms of the weight of the cannon ball. Thus, an artillery gun would be designated a '12-pounder', a '16-pounder', and so on.

A similar concept is preserved for designating shotgun calibre in terms of its 'bore' ('gauge' in the United States). A 12-bore or 12-gauge shotgun originally meant that it fired balls that weighed 12 to the pound. This meant that a 16-bore shotgun, despite having a higher bore number, fired a smaller shot and had a smaller calibre than a 12-bore. The smaller the bore number, the larger the bore or calibre. Nowadays, however, with shotgun bullets no longer being spherical, the direct relationship between bore and weight of bullet no longer holds and shotgun calibres are specified like other guns as above.

| Bore | inches | millimetres |
|------|--------|-------------|
| 8 | 0.835–0.860 | 21.2–21.8 |
| 10 | 0.775–0.793 | 19.6–20.1 |
| 12 | 0.729–0.740 | 18.5–18.8 |
| 16 | 0.662–0.669 | 16.8–17.0 |
| 18 | 0.637 | 16.2 |
| 20 | 0.615 | 15.6 |
| 24 | 0.579 | 14.7 |
| 28 | 0.550 | 14.0 |
| 32 | 0.526 | 13.4 |

Other sporting guns, as well as pistols and revolvers, have familiar calibres measured in inches, such as the .22 – usually called a 'two-two' in the UK ('twenty-two' in the US) – or in millimetres in Continental Europe. In the armed forces, since the advent of Nato in 1949, calibre stated in millimetres has become standard in nearly all western nations. However, larger-calibre guns are still sometimes referred to in both inches and millimetres, for example a '4-inch mortar'.

Strictly speaking, calibre is not the internal diameter of the barrels of these guns, since they are generally 'rifled' – in other words there is a spiral grooving that grips the bullet and spins it to ensure stability. The barrel therefore has a cross-section like this:

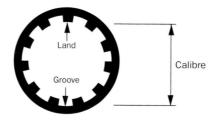

that consists of grooves and so-called 'lands'. The calibre is given by the smaller interior diameter of the barrel, between the lands.

Above: **Callipers for measuring cannon balls. They were made from bronze by Paul Revere of Boston, the American revolutionary leader.**

Above: **'Magnum 44' revolver, made by Smith & Wesson. It is the chosen weapon of Harry Callahan, the tough-cop hero of *Dirty Harry*. Its calibre is 0.44 in.**

# Economics

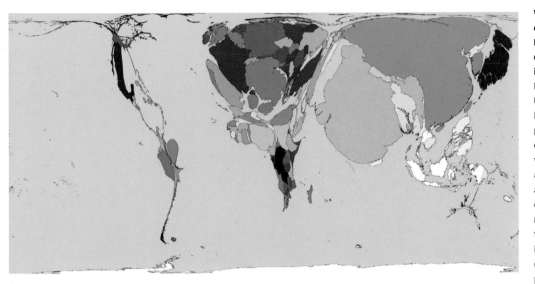

The worlds of business and economics are awash with numbers, measuring every aspect of wealth from salaries, company stock and share prices to the human development index of the United Nations and gross national product. But for all its aspirations to be a quantitative science, economics is constantly undermined by the realities of business.

Scientists, though often competitive, have a shared culture of trust in their published results; businessmen have to be competitive and treat company reports with scepticism. In science, art and judgement are required in order to fit observational data to natural laws; in business, they are needed to massage financial figures to fit human laws and expectations. Theodore Porter reminds us in *Trust in Numbers*: 'heroic entrepreneurship and criminal embezzlement may be distinguished by no more than a subtle point enunciated… by the regulatory agencies.'

The most familiar economic indicators are probably the New York-based Dow-Jones industrial average and the London-based *Financial Times*-Stock Exchange 100 share index (FT-SE or 'Footsie' 100). The Dow-Jones index appeared in 1897 from a financial news publisher founded by Charles Dow and Edward Jones, and was an arithmetical average calculated daily by dividing the total price of a list of 12 stocks by 12. Since 1928, the most commonly quoted Dow-Jones index has been a daily average of the prices of 30 industrial stocks accounting for about 25 per cent of the market, adjusted to compensate for stock splits, stock substitutions and significant dividend changes. The FT-SE 100 replaced the FT 30 index, introduced in 1935, and is based on the market capitalization of the 100 largest industrial and commercial companies quoted on the London Stock Exchange, which are reviewed quarterly. It is a weighted average and is calculated minute by minute, from a base index chosen to be 1000 on 3 January 1984.

Wealth in 1500. The area of each country in this map is proportional to its gross domestic product (GDP) in 1500, as measured in Purchasing Power Parity US$ at 1990 prices (thus PPP US$1 has the same purchasing power in every country). GDP is the total value of goods produced and services provided in a country in one year; it excludes the value of raw materials used in producing these goods, and net income from abroad (included in gross national product, GNP). If GDP is divided by a country's population in 1500, Italy (almost unrecognizably bloated) had by far the highest wealth per person at $1100, with the UK at $714, India at $550 and Egypt at $475; North America and Australia barely register. The map is one of dozens created by the Worldmapper project using an algorithm developed by Michael Gastner and Mark Newman.

# Lotteries and Gambling

The billionaire cricket promoter Kerry Packer was a compulsive gambler, who once lost several million dollars at a Las Vegas casino. Publicly criticized for his profligate lifestyle, Packer was unrepentant: 'My father was a gambler, I am a gambler. Every man who ever created anything was a gambler.'

Gambling is even older than money. It appeals across the spectrum, though historically more to the aristocracy and the lower orders than the middle classes. In the South Sea Bubble of 1720, King George I and Sir Isaac Newton both gambled and lost. Smaller fry invested in simultaneous fly-by-night schemes. An unknown adventurer even advertised '*A company for carrying on an undertaking of great advantage; but nobody to know what it is.*' Five hours after opening an office, he had sold 1,000 shares for a deposit of £2 each. Then, 'philosopher enough to be contented with his venture', he shut up shop and set off for the Continent that evening. 'Were not the fact stated by scores of credible witnesses, it would be impossible to believe that any person could have been duped by such a project', wrote Charles Mackay in his classic *Extraordinary Popular Delusions and the Madness of Crowds*.

The computation of chance was first systematized by the Italian physician, mathematician, astrologer and ferocious gambler Geronimo Cardano, who published *Liber de Ludo Aleae* (Book on Games of Chance) in 1525. Cardano was followed a century later by Blaise Pascal and Pierre de Fermat, the French mathematicians who founded the theory of probabilities.

But of course most gamblers believe in luck – especially those who play national lotteries for a few fantastically big prizes, where the odds are millions to one against success. Lotteries were widespread in Europe by the 16th century and had papal approval, judging from a Parisian lottery of 1572, in which the winning tickets carried the phrase 'God has chosen you' and the losing ones 'God comforts you'. They became major sources of revenue. The construction of London's water supply, Westminster Bridge and the British Museum all had lottery funding. After a gap from 1826 to 1994, when a national lottery was frowned upon, lottery funding is again important in Britain for many charitable, educational and artistic activities. 'What lottery providers have known for centuries is that you can get somebody to pay for a one-in-a-million shot, more than the value of that chance. In other words, people pay more for a claim on a very big pay-off', said the chairman of the US Federal Reserve Alan Greenspan at the time of the dotcom bubble in 2000.

Above: **Geronimo Cardano (1501–76), who first computed probabilities in gambling. He regarded gambling, including his own, as an addiction requiring medical treatment. He advised: 'The greatest advantage in gambling comes from not playing at all.'**

Below: **The lottery with the largest prize in Europe is the Christmas draw of the Spanish National Lottery, known as 'El Gordo' (the Fat One). In 2004 it was won by a man from the town of Sort – which in the local language means 'luck'.**

# Sports and Games

Think of a sport like football, cricket or tennis – and movement and physical skill immediately come to mind. Think of a game like chess, bridge or Scrabble – and the image is one of stillness and mental exercise. Yet, football, cricket and tennis are often known as games as well as sports, though the reverse is not true for chess, bridge and Scrabble. Whether sports or games, these are all competitive and require regulations and rules – a key form of measurement. The traditional distinction between sports and games, based on criteria such as whether an activity had codified rules or not, whether it involved a team or individuals, and whether it took place in the outdoors or in a prescribed or enclosed space, has become muddied, and the two words are now often used interchangeably.

Intuition tells us that sports and games must have existed in early societies for reasons similar to those in our modern ones: they were good ways to socialise, to keep physically fit, to sharpen wits, to show off prowess and to entertain and be entertained. However, there is no unambiguous archaeological evidence for competitive sports and games until the early 3rd millennium BC, when Mesopotamian cuneiform tablets contain brief references to wrestling or boxing bouts with defined rules.

The traditional date for the founding of the Olympic Games is 776 BC; and its modern revival dates from 1896. The sole recorded event from that early date is a sprint known as a *stade* race of about 200 metres, or one length of the original stadium at Olympia – won by Koroibos of Elis. But the ancient games grew to include a longer-distance race, wrestling, boxing, free fighting and a pentathlon, spread over four days.

Renaissance sports preserved the aesthetic element found in ancient Greek

Left: **The Discobolus. It shows a discus thrower about to release his throw. Although this statue, which is 1.55 metres in height, has long been considered the personification of ancient Greek athleticism, its stance is today considered a rather inefficient way to throw a discus – as indicated by the modern discus thrower Sukhbir Singh** (above), **at the South Asian Federation Games in 2004. The statue is a Roman copy in marble, found in Rome in 1781, based on a lost Greek original in bronze, sculpted by Myron in the period 460–450 BC. The original probably commemorated a winner of the pentathlon in one of the major Greek games of the 5th century.**

sport. But with the coming of the scientific revolution, aesthetics increasingly gave way to quantification, although it survives today in figure skating, diving and gymnastics. Sports equipment improved, as did the capacity to measure an athlete's performance; new games, such as basketball, were consciously invented to take advantage of this change. The concept of the sports record appeared. While the word itself, meaning an unsurpassed quantified achievement, dates from the late 19th century, first in English, the concept is about two centuries old.

Today, wherever one looks in sport, people are measuring, whether it be the time of a race, the number of goals, the dimensions of a swimming pool or the distance of a javelin throw. The difference between winning and losing can sometimes be just a hundredth of

a second, requiring a computer to measure it. 'Maybe we are too fanatical about it', notes the Australian government's National Standards Commission, 'but, then again, there are few things more frustrating than an inconclusive result.'

**Board games and cards.** Board games are much older than cards. Chess, shown left in Satyajit Ray's film *The Chess Players* (1977), was invented in India in the 1st millennium AD; it came via Persia to Europe, where the rules were slightly modified to make the game go faster. Cards, shown above in Georges de la Tour's *The Card Sharp, c.* 1620–40, existed in Europe from around the 12th century. Games and cheating have always gone together: in the painting, the man is hiding the ace of clubs in his cummerbund, and in the film, one chess player later moves a piece while the other is out of the room – even though, unlike the card players, the chess players are not playing for money but simply for love of the game.

# Collective Nouns

So many people are fascinated by the names for groups of animals and things that the *Oxford English Dictionary* has a weblink dedicated to the subject. Cows come in herds, flies in swarms and rags in bundles – but what about bears, foxes and ladies? In 1486, the *Book of St Albans* noted that a strict regard for these collective niceties distinguished 'gentylmen from ungentylmen' better than knowledge of the rules of grammar: 'We say a congregacyon of people, a hoost of men, a felyshyppynge of yeomen, and a bevy of ladyes; we must speak of a herde of deer, swannys, cranys, or wrenys, a sege of herons or bytourys [bitterns], a muster of pecockes, a watche of nyghtyngales, a flyghte of doves, a claterynge of choughes, a pryde of lyons, a slewthe of beeres [bears], a gagle of geys, a skulke of foxes, a sculle of frerys [friars], a pontificalitye of prestys [priests], and a superfluyte of nonnes [nuns].' The last three were probably ironic.

The nouns in this list are selected from *Brewer's Dictionary of Phrase and Fable* (17th edn), and cross-checked against the *Oxford English Dictionary*. Many are in common use, some are rare, and a few (such as an 'exaltation' of larks) have been revived. Their origins are often delightfully clear: the noun expresses the key characteristic of the group. Others are puzzling. For cats, both *Brewer's* and the *OED* weblink recommend 'clowder'. But 'clowder' does not appear in the *Shorter Oxford English Dictionary* and is certainly not common; indeed, the very idea of cats assembling in a group seems odd.

| | |
|---|---|
| angels | a host |
| arrows | a sheaf |
| badgers | a cete |
| bears | a sloth |
| bells | a peal |
| boars | a sounder |
| cars | a fleet |
| cattle | a drove, a herd |
| cats | a clowder |
| chickens | a brood |
| choughs | a chattering |
| crows | a murder |
| cubs | a litter |
| dolphins | a school |
| eggs | a clutch |
| ferrets | a business, a fesnying |
| finches | a charm or chirm |
| foxes | a skulk |
| geese | (in flight) a skein; (on the ground) a gaggle |
| girls | a bevy |
| grouse | a covey |
| gulls | a colony |
| hares | a huske |
| herons | a siege |
| kittens | a kindle |
| labourers | a gang |
| larks | an exaltation |
| lions | a pride |
| magistrates | a bench |
| mares | a stud |
| moles | a labour |
| monkeys | a troop |
| mules | a barren |
| nightingales | a watch |
| onions | a rope, a string |
| owls | a parliament |
| peacocks | a muster |
| ravens | an unkindness |
| rooks | a building, a clamour |
| sails | a suit |
| savages | a horde |
| snipe | a wisp |
| starlings | a murmuration |
| stars | a cluster, a constellation |
| teal | a spring |
| whales | a school, a gam, a pod |
| wolves | a pack, a rout, a herd |

'A galaxy of astronomers', from *An Exaltation of Larks*, a book on collective nouns by James Lipton.

# The Measure of All Things

'Man is the measure of all things', said the philosopher Protagoras, 2,500 years ago. 'We must remember that measures were made for man and not man for measures', said Isaac Asimov, the scientist and science-fiction writer, in our own time.

When we look at an image of vast, unknown galaxies spinning in space at the furthest edge of the visible world, made by the Hubble space telescope – such as the image on p. 80 – it is difficult to agree with thinkers like Protagoras and Asimov. How can the Hubble image *not* induce in us a feeling that humankind is the measure of almost nothing? Surely, what we are is a handful of insignificant specks of matter in a totally indifferent, seemingly infinite Universe. And yet, and yet. Those Hubble images were fabricated by human beings, using amazingly sophisticated technology also made by human beings, whose minds were capable of making astonishing invisible calculations with such concepts as billions of light-years, black holes and the Big Bang. Recall Archimedes, with his Sand Reckoner, who estimated the size of the Universe by mentally filling it with grains of sand, as mentioned in the introduction. In some ways, the human mind is an even more incredible thing to contemplate than the objective Universe 'out there' in space or within each atomic nucleus. As the director of *2001: A Space Odyssey*, Stanley Kubrick, said of the film's writer, Arthur C. Clarke: 'Arthur somehow manages to capture the hopeless but admirable human desire to know things that can really never be known.'

When Albert Einstein met the writer, artist, musician, philosopher and fellow Nobel laureate Rabindranath Tagore in 1930, they had several fascinating conversations on the issue addressed by Protagoras and Asimov, and they profoundly disagreed with each other. One of the conversations was published in *The New York Times* with a photograph of Einstein and Tagore together and the teasing caption: 'A Mathematician and a Mystic Meet in Manhattan'. Einstein said: 'There are two different conceptions about the nature of the Universe – the world as a unity dependent on humanity, and the world as reality independent of the human factor…' Einstein made it plain that he believed in the second conception. Tagore replied: 'This world is a human world – the scientific view of it is also that of the scientific man. Therefore, the world apart from us does not exist; it is a relative world, depending for its reality upon our consciousness.' Clearly, Tagore believed in the first conception of the Universe.

Of course Protagoras may have intended his statement to be taken at several levels, from the sublimely philosophical to the mundanely commercial. Already, in 5th-century BC Greece, and for a few thousand years before this, practical men had been measuring the world according to the parts of the human body – fingers, feet, arms and so on. In the subsequent centuries of the Roman Empire and the Middle Ages, human-scale dimensions continued to dominate measurement – via such units as the mile, the acre, the bushel and the pound.

Opposite: **Taking the measure of the Moon. This pristine footprint in the Moon's surface was made by the boot of Buzz Aldrin, the second man on the Moon, during the Apollo 11 mission in 1969. Given that an estimated 10 million years are required for the constant rain of micro-meteorites to churn the uppermost half-inch of lunar soil, the footprint should stay undisturbed – unless, as the astronomer Patrick Moore observes, 'it is taken away to a lunar museum.' We can even imagine that if human beings as a species are reckless enough, the footprint will remain, like a sort of fossil, long after the last humans have extinguished themselves from the Earth.**

*Fool's Cap World, c.* 1590. 'This startling and disturbing image is one of the enigmas of cartographic history', writes Peter Whitfield in *The Image of the World: 20 Centuries of World Maps.* 'The artist, date and place of publication are all unknown, and its purpose can only be guessed at.' The artist gives a Latin assumed name in the panel on the left: Epichthonius Cosmopolites (roughly speaking, 'Everyman'). Since the map's geography closely resembles Abraham Ortelius's world maps of the 1580s, and there is one reference to the image in Robert Burton's *Anatomy of Melancholy* (1621), the image carries the tentative date 1590 – the period of Shakespeare's early plays, in which the figure of the Fool has a vital role. What does it mean? 'Its central visual metaphor is the universality of human folly.' But further interpretations are clearly possible, not least to do with 21st-century preoccupations with global environmental catastrophe. In the context of man and measurement, the image is a visual equivalent of Protagoras's ambiguous statement, 'Man is the measure of all things' – packed with meaning and irony.

Only with the coming of the scientific mindset in the 17th century, the metric system, and the drive for precision in the 19th century, did measures increasingly become based on nature rather than man. 'For about two centuries, quantitative precision has been understood as central to experimental science,' notes Theodore Porter in *Trust in Numbers*. 'Precision has been valued as a sign of diligence, skill and impersonality.' The key word here is impersonality: the more that measures are divorced from man, the more scientific they are taken to be.

But the strange – and at times amusing – fact is that scientists, for all their reverence for impersonality in science, are actually passionate personalizers of science and measurement. Forget the Nobel prize: the greatest honour in science is for a scientific law, such as Boyle's law, or a unit of measurement, such as the watt, to be named after a person. Of the seven basic units in the Système International, the ampere and the kelvin are named after scientists. Among the many derived SI units, scientists' names easily predominate, for example, the hertz (for frequency), the newton (for force), the pascal (for pressure), the joule (for energy), the volt (for electrical potential), the ohm (for electrical resistance), the farad (for electrical capacitance), the watt (for power), the becquerel (for radioactivity), the degree Celsius (for temperature), and others.

The introduction of new units naturally provokes controversy. This is partly because scientists rightly do not want to complicate the existing SI units, but also because a name publicly apportions credit for work, and not everyone can agree on who deserves the credit. In the 1920s, the German physics establishment proposed a new unit for frequency, then referred to in cycles per second, to be named after Heinrich Hertz. A well-known German physical chemist, Walther Nernst, remarked acidly: 'I do not see the necessity of introducing a new name; by the same reasoning one might as well call 1 litre per second 1 'falstaff'!' In a more kindly joking spirit, some students at Cambridge University proposed a new unit named after one of their lecturers, P. A. M. Dirac, a legendary theoretical physicist (who won a Nobel prize aged 31). Dirac was famously unwilling to speak, and so the dirac was to be the unit of prevalence of silence during discourse. It is nicely defined by Arthur Klein in *The World of Measurements*: 'Just as the ohm of electrical resistance measures the opposition to current flow in response to an electromotive force, so the dirac suggested reluctance to speak unless speaking is unavoidable.'

Scientists' passionate interest in the naming of units is readily understandable, given that measurement lies at the root of science. It is worth repeating Lord Kelvin's celebrated comment – which is set in stone on the *social* sciences building at the University of Chicago – 'when you can measure what you are speaking about and express it in numbers you know something about it, but when you cannot measure it in numbers, your knowledge is of a meagre and unsatisfactory kind.' But it does not follow from this that just because you can measure something, you *should* measure it – or that measurements have an intrinsic value because they are precise.

To descend for a moment from the sublime and the mundane to the ridiculous, consider the American physician William Bean, who measured the growth of his thumbnail every day from the age of 32 onwards, by filing a horizontal line above the cuticle and then recording its growth towards the tip of his thumb. This study culminated in a scientific paper entitled 'Nail growth: 35 years of observation'. Its main conclusion was that, regardless of where he was, Bean's nails grew at the same steady rate: approximately a tenth of a millimetre per day.

It is easy to scoff at such pointless (if inexpensive) research – and in recent years many have done so, with the launching in 1991 of the Ig Nobel prizes for scientific discoveries 'that cannot, or should not, be reproduced'. The aim of these annual awards for published scientific work is: 'first make people laugh, and then make them think'. Perhaps the Ig Nobel prizes have slightly diminished the respect for genuinely worthwhile, if recondite scientific research. But overall they are a good antidote to measurement mania. The basic reason we laugh at the Ig Nobels must be that we recognize that measurement is in danger of running amok in the modern world – and not just in science. Accountants and consultants and economists are stifling companies; multiple-choice questions and journal impact factors and league tables are stifling education; pollsters and focus groups and even lie detectors are stifling government. The audit society, with its numbers game and bean-counting, threatens to strangle the very activities it is intended to measure. After tolerating it for a while, both the measured and the measurer become demoralized by the mediocrity, conformism and waste that attend such dehumanizing faith in statistics, targets and money. In many organizations and institutions, alas, humans are now made for measuring, rather than the reverse called for by Asimov.

So here is a third level at which Protagoras's statement has meaning, located somewhere between the philosophical and the mundane. Perhaps we should rephrase him in this respect as: 'Humanity is the measure of all things.'

It is wonderful how thought-provoking those ancient Greeks still are. But the story of measurement told in this book demonstrates that humankind has moved way beyond the ancient Greeks in science, if not in psychology, economics and politics. If we welcome the modern world, then we have no choice but to embrace the experimental observation of nature, the necessity for precision and the development of ever-more magical technology. Before science, man may have been the measure of all things. Today, man and measurement strike me as being more like longstanding partners in a devoted but headstrong marriage. Man plus measurement equals science. The challenge is how to make this equation balance.

# Further Reading

This is not a scholarly bibliography but rather a selection of books and articles directly relevant to each chapter of this book. Only books that deal substantially with measurement are included. The date given is usually the date of first publication in English, unless otherwise indicated. Online references are not included, but I must recommend www.wikipedia.org as an invaluable resource, if used with discrimination, and the useful website of the National Physical Laboratory, www.npl.co.uk, especially the NPL's 'Beginner's guides to measurement' and its newsletter *Metromnia*. The journals *Metrologia*, *Nature* and *Science*, and the magazines *Geographical*, *New Scientist*, *Physics World* and *Scientific American*, are valuable sources on current methods of measurement. An excellent source of historical and biographical information is *The Hutchinson Dictionary of Scientific Biography*, vols 1 and 2, 3rd edn, 2000 (Roy Porter and Marilyn Ogilvie, consultant eds).

## Introduction/General Works
Darton, Mike and John Clark, *The Dent Dictionary of Measurement*, 1994
Hebra, Alex, *Measure for Measure: The Story of Imperial, Metric, and Other Units*, 2003
Klein, H. Arthur, *The World of Measurements: Masterpieces, Mysteries, and Muddles of Metrology*, 1974
Kula, Witold, *Measures and Men*, 1986
Morrison, Philip and Phyllis Morrison and The Office of Charles and Ray Eames, *Powers of Ten: About the Relative Size of Things in the Universe*, 1982
*Nature*, 'Small scale', 27 Apr. 2006: 1092 (on the zepto-world)
Porter, Theodore M., *Trust in Numbers: The Pursuit of Objectivity in Science and Public Life*, 1995
Robinson, Andrew, *The Last Man Who Knew Everything: Thomas Young*, 2006
Tufte, Edward R., *The Visual Display of Quantitative Information*, 1983
Young, Thomas, 'On weights and measures', in *Miscellaneous Works of the Late Thomas Young*, (George Peacock, ed.), vol. 2, 2003

## Going Metric
Alder, Ken, *The Measure of All Things: The Seven-Year Odyssey That Transformed the World*, 2002
Barber, Peter, ed., *The Map Book*, 2005
Berthon, Simon and Andrew Robinson, *The Shape of the World: The Mapping and Discovery of the Earth*, 1989
Danson, Edwin, *Weighing the World: The Quest to Measure the Earth*, 2006
Gillispie, Charles Coulston, *Science and Polity in France: The Revolutionary and Napoleonic Years*, 2004
Sobel, Dava and William J. H. Andrewes, *The Illustrated Longitude*, 1998
UK Metric Association, *A Very British Mess*, 2004

and *Metric Signs Ahead*, 2006 (reports)
Westfall, Richard S., *The Life of Isaac Newton*, 1993
Wilford, John Noble, *The Mapmakers: The Story of the Great Pioneers in Cartography from Antiquity to the Space Age*, 1981
Zupko, Ronald Edward, *Revolution in Measurement: Western European Weights and Measures since the Age of Science*, 1990

## Number and Mathematics
Barrow, John D., *The Infinite Book: A Short Guide to the Boundless, Timeless and Endless*, 2005
Cohen, I. B., *The Triumph of Numbers: How Counting Shaped Modern Life*, 2005
Dantzig, Tobias, *Number: The Language of Science*, new edn, 2006
Dilke, O. A. W., *Mathematics and Measurement*, 1987
Einstein, Albert, 'Geometry and experience' and 'On the method of theoretical physics', in Einstein, *Ideas and Opinions*, 1954
Hodgkin, Luke, *A History of Mathematics: From Mesopotamia to Modernity*, 2005
Livio, Mario, *The Golden Ratio: The Story of Phi, the World's Most Astonishing Number*, 2003
Mandelbrot, Benoit B., *The Fractal Geometry of Nature*, rev. edn, 1983
— 'Fractals as a morphology of the amorphous', in Bill Hirst, *Fractal Landscapes from the Real World*, 1994
Quilter, Jeffrey and Gary Urton, *Narrative Threads: Accounting and Recounting in Andean Khipu*, 2002
Seife, Charles, *Zero: The Biography of a Dangerous Idea*, 2000
Stoll, Cliff, 'When slide rules ruled', *Scientific American*, May 2006
Taylor, Richard P., 'Order in Pollock's chaos', *Scientific American*, Dec. 2002
Wigner, Eugene, 'The unreasonable effectiveness of mathematics in the natural sciences', *Communications in Pure and Applied Mathematics*, Feb. 1960

## Customary Units
Abbott, Alison, 'Rebuilding the past', *Nature*, 16 Dec. 2004: 794-95,(on the elephant water clock)
Battersby, Stephen, 'The lady who sold time', *New Scientist*, 25 Feb. 2006: 52-53 (on Ruth Belville)
Berriman, A. E., *Historical Metrology: A New Analysis of the Archaeological and the Historical Evidence Relating to Weights and Measures*, 1953
Chapman, Allan, *Dividing the Circle: The Development of Critical Angular Measurement in Astronomy 1500-1850*, 2nd edn, 1995
Collins, Paul, 'The sweet sound of profit', *New Scientist*, 20 May 2006: 54-55 (on cash registers)
Connor, R. D., *The Weights and Measures of England*, 1987
Goetzmann, William N. and K. Geert Rouwenhorst, eds, *The Origins of Value: The Financial Innovations that Created Modern Capital Markets*, 2005

Holford-Stevens, Leofranc, *The History of Time: A Very Short Introduction*, 2005
Kenoyer, Jonathan Mark, *Ancient Cities of the Indus Valley Civilization*, 1998
Prerau, David, *Saving the Daylight: Why We Put the Clocks Forward*, 2005
Richardson, W. F., *Numbering and Measuring in the Classical World*, rev. edn, 2004
Shaw, Ian and Paul Nicholson, *British Museum Dictionary of Ancient Egypt*, 1995

## Instruments and Techniques
Ackland, Len, 'Radiation: how safe is safe?', *New Scientist*, 15 May 1993: 34-37
Anderson, Katherine, *Predicting the Weather: Victorians and the Science of Meteorology*, 2005
Chang, Hasok, *Inventing Temperature: Measurement and Scientific Progress*, 2004
Gilmozzi, Roberto, 'Giant telescopes of the future', *Scientific American*, May 2006
Hecht, Jeff, *Beam: The Race to Make the Laser*, 2005
Hogan, Jenny, 'Focus on the living', *Nature*, 2 Mar. 2006: 14-15 (on the atomic force microscope)
Horsfall, Alton and Nick Wright, 'Sensing the extreme', *Physics World*, May 2006 (on high-temperature measurement)
Magnello, Eileen, *A Century of Measurement: An Illustrated History of the National Physical Laboratory*, 2000
Musher, Daniel M., Edward A. Dominguez and Ariel Bar-Sela, 'Edward Seguin and the social power of thermometry', *New England Journal of Medicine*, 8 Jan. 1987: 115-17
Nellist, Peter, 'Seeing with electrons', *Physics World*, Nov. 2005 (on microscopes)
Peplow, Mark, 'Counting the dead', *Nature*, 20 Apr. 2006: 982-83 (on Chernobyl)

## Atoms
Ancey, Christophe and Steve Cochard, 'Understanding avalanches', *Physics World*, July 2006
Atkins, P. W., *The 2nd Law: Energy, Chaos, and Form*, 1984
— *The Periodic Kingdom: A Journey into the Land of the Chemical Elements*, 1995
Ball, Philip, *The Elements: A Very Short Introduction*, 2002
Bowman, Sheridan, *Radiocarbon Dating*, 1990
Einstein, Albert, *Relativity: The Special and the General Theory*, Routledge Classics edn, 2001
Einstein, Albert and Leopold Infeld, *The Evolution of Physics*, 1938
Lovelock, James, *Homage to Gaia: The Life of an Independent Scientist*, 2000
Rigden, John S., *Einstein 1905: The Standard of Greatness*, 2005
Robinson, Andrew, *Einstein: A Hundred Years of Relativity*, 2005
Robinson, Ian, 'Redefining the kilogram', *Physics World*, May 2004

Rothemund, Paul W. K., 'Folding DNA to create nanoscale shapes and patterns', *Nature*, 16 Mar. 2006: 297-302

Smith, Lloyd M., 'The manifold faces of DNA', *Nature*, 16 Mar. 2006: 283-84

Sykes, Christopher, *No Ordinary Genius: The Illustrated Richard Feynman*, 1994

## Earth

Amodeo, Christian, 'Eyes in the skies', *Geographical*, Feb. 2003

Bluestein, Howard B., *Tornado Alley: Monster Storms of the Great Plains*, 1999

Choi, Charles, 'Volcanic sniffing', *Scientific American*, Nov. 2004

Dwyer, Joseph R., 'A bolt out of the blue', *Scientific American*, May 2005

Emanuel, Kerry, *The History and Science of Hurricanes*, 2005

Fara, Patricia, *Fatal Attraction: Magnetic Mysteries of the Enlightenment*, 2005

Flindt, Rainer, *Amazing Numbers in Biology*, 2006

Gradstein, Felix, James Ogg and Alan Smith, eds, *A Geologic Time Scale 2004*, 2004

Hough, Susan Elizabeth, *Richter's Scale: Measure of an Earthquake: Measure of a Man*, 2007

Jackson, Patrick Wyse, *The Chronologers' Quest: The Search for the Age of the Earth*, 2006

Keay, John, *The Great Arc: The Dramatic Tale of How India Was Mapped and Everest Was Named*, 2000

Lawson, Simon, 'Spotting a fake', *Physics World*, June 2006 (on diamonds)

Lopes, Rosaly, *The Volcano Adventure Guide*, 2005

Nouvian, Claire, *The Deep: The Extraordinary Creatures of the Abyss*, 2007

Pavord, Anna, *The Naming of Names: The Search for Order in the World of Plants*, 2005

Pretor-Pinney, Gavin, *The Cloudspotter's Guide*, 2006

Robinson, Andrew, *Earthshock: Hurricanes, Volcanoes, Earthquakes, Tornadoes and Other Forces of Nature*, rev. edn, 2002

Rudwick, Martin J. S., *Bursting the Limits of Time: The Reconstruction of Geohistory in the Age of Revolution*, 2005

Scarpa, Roberto, 'Predicting volcanic eruptions', *Science*, 27 July 2001: 615-16

Schmincke, Hans-Ulrich, *Volcanism*, 2004

Schrope, Mark, 'The bolt catchers', *Nature*, 9 Sept. 2004: 120-21 (on lightning)

Titov, Vasily *et al.*, 'The global reach of the 26 December 2004 Sumatra tsunami', *Science*, 23 Sept. 2005: 2045-48

Wilson, Edward O., *In Search of Nature*, 1996

Ziebart, Marek, 'The story of height', *Geographical*, Aug. 2003

Zijlstra, Albert, 'The last word', *New Scientist*, 8 Nov. 2003 (on clouds)

## Universe

Barrow, John D. and John K. Webb, 'Inconstant constants', *Scientific American*, June 2005

Bonnell, Jerry T. and Robert J. Nemiroff, *Astronomy: 365 Days*, 2006

Chandler, David L., 'It's time to go back', *New Scientist*, 1 Apr. 2006: 32-37 (on science on the Moon)

Christensen, Lars Lindberg and Bob Fosbury, *Hubble: 15 Years of Discovery*, 2006

Davies, Paul, *The Goldilocks Enigma: Why is the Universe Just Right for Life?*, 2006

Einstein, Albert, 'Johannes Kepler', in Einstein, *Ideas and Opinions*, 1954

Hill, Steele and Michael Carlowicz, *The Sun*, 2006

Hinshaw, Gary, 'WMAP data put cosmic inflation to the test', *Physics World*, May 2006

Holman, Gordon D., 'The mysterious origins of solar flares', *Scientific American*, Apr. 2006

Light, Michael, *Full Moon*, 2006

Lovett, Laura, Joan Horvath and Jeff Cuzzi, *Saturn: A New View*, 2006

Moore, Patrick, *Patrick Moore on the Moon*, 2001

*Nature*, 'Neglected neighbour', 13 Apr. 2006: 846 (editorial on Venus versus Mars)

Peplow, Mark, 'Comet chasers get mineral shock', *Nature*, 16 Mar. 2006: 260

Singh, Simon, *Big Bang*, 2004

Weinberg, Steven, *Dreams of a Final Theory: The Search for the Fundamental Laws of Nature*, 1993

## Mind

Ball, Philip, 'Index aims for fair ranking of scientists', *Nature*, 18 Aug. 2005: 900 (on the h-index)

— 'Prestige is factored into journal rankings', *Nature*, 16 Feb. 2006: 770–71

DeFrancis, John, *Visible Speech: The Diverse Oneness of Writing Systems*, 1989

Domino, George and Marla L. Domino, *Psychological Testing: An Introduction*, 2nd edn, 2006

Gould, Stephen Jay, *The Mismeasure of Man*, rev. edn, 1996

Howard, David, 'Staying in tune with physics', *Physics World*, Apr. 2006 (on singing)

Lenman, Robin, ed., *The Oxford Companion to the Photograph*, 2005

Loxley, Simon, *Type: The Secret History of Letters*, 2004

Mitchell, Michael and Susan Wightman, *Book Typography: A Designer's Manual*, 2005

*Nature*, 'Not-so-deep impact', 23 June 2005: 1003-04 (editorial on impact factors)

Naughton, John, *A Brief History of the Future: The Origins of the Internet*, 1999

Petroski, Henry, *The Evolution of Useful Things*, 1992

Pinker, Steven, *The Language Instinct: The New Science of Language and Mind*, 1994

Robbins Landon, H. C., *Mozart: The Golden Years*, 1989

Robinson, Andrew, *The Man Who Deciphered Linear B: The Story of Michael Ventris*, 2002

— *Lost Languages: The Enigma of the World's Undeciphered Scripts*, 2002

— *The Story of Writing: Alphabets, Hieroglyphs and Pictograms*, 2nd edn, 2007

Shaw, P. *et al.*, 'Intellectual ability and cortical development in children and adolescents', *Nature*, 30 Mar. 2006: 676-79

Taylor, Arlene G., *Wynar's Introduction to Cataloging and Classification*, 9th edn, 2000

White, John, *Intelligence, Destiny and Education: The Ideological Roots of Intelligence Testing*, 2006

## Body

Blakemore, Colin and Sheila Jennett, eds, *The Oxford Companion to the Body*, 2001

Bond, Shirley, *Home Measures: The Essential Reference Guide to Sizes and Measurements for Home, Office and Kitchen*, 1996

Gibbs, W. Wayt, 'Obesity: an overblown epidemic?', *Scientific American*, June 2005

Goodson, Boyd, 'Mobilizing magnetic resonance', *Physics World*, May 2006

Gosline, Anna, 'Will DNA profiling fuel prejudice?' *New Scientist*, 9 Apr. 2005: 12-13

Hempel, Sandra, *The Medical Detective: John Snow and the Mystery of Cholera*, 2006

Kemp, Martin, *Leonardo da Vinci: The Marvellous Works of Nature and Man*, 2006

Melzack, Ronald and Patrick D. Wall, *The Challenge of Pain*, 2nd edn, 1996

*Nature*, 'Coping with complexity', 25 May 2006: 383–4 (editorial on the nature of the gene)

Naylor, G. R. S., 'A simple togmeter for measuring the warmth of continental quilts', 1994 (report at www.tft.csiro.au)

Pain, Stephanie, 'Davy's dark side', *New Scientist*, 3 Sept. 2005: 48-49

Pope, Jean, *Medical Physics: Imaging*, 1999

Porter, Roy, *The Greatest Benefit to Mankind: A Medical History of Mankind from Antiquity to the Present*, 1997

Shapin, Steven, 'Eat and run: why we're so fat', *New Yorker*, 16 Jan. 2006: 76-82

Vince, Gaia, 'The many ages of man', *New Scientist*, 17 June 2006: 50–53

Watson, James D., *DNA: The Secret of Life*, 2003

Westphal, Sylvia Pagán, 'Red alert', *New Scientist*, 23 July 2005: 33–6 (on blood)

## Society

Atherton, Mike, *Gambling: A Story of Triumph and Disaster*, 2006

Bamshad, Michael J. and Steve E. Olson, 'Does race exist?', *Scientific American*, Dec. 2003

Check, Erica, 'The tiger's retreat', *Nature*, 22 June 2006: 927-30

Cole, Simon A., 'Misplaced convictions', *New Scientist*, 18 Mar. 2006: 23 (on fingerprint evidence)

# Illustration Credits

Davies, Simon, 'Iris recognition', *New Scientist*, 20/27 Dec. 2004: 34 (letter)

Doyle, Rodger, 'Calculus of happiness', *Scientific American*, Nov. 2002

— 'Religion in America', *Scientific American*, Feb. 2003

Grossman, Wendy M., 'Ballot breakdown', *Scientific American*, Feb. 2004 (on electronic voting)

Jerome, Fred and Rodger Taylor, *Einstein on Race and Racism*, 2005

Kevles, Daniel J., 'Grounds for breeding', *Times Literary Supplement*, 2 Jan. 1998 (on eugenics)

Lipton, James, *An Exaltation of Larks*, 3rd edn, 1991

Lykken, David, 'Nothing like the truth', *New Scientist*, 14 Aug. 2004: 17 (on lie detectors)

Mackay, Charles, *Extraordinary Popular Delusions and the Madness of Crowds*, 1841

Miller, Shaun and Jared Diamond, 'A New World of differences', *Nature*, 25 May 2006: 411-12 (on GDP)

Pearson, Helen, 'Lure of lie detectors spooks ethicists', *Nature*, 22 June 2006: 918-19

Poole, Robert, 'Making up for lost time', *History Today*, Dec. 1999

Randerson, James and Andy Coughlan, 'Forensic evidence stands accused', *New Scientist*, 31 Jan. 2004: 6-7

Raper, J. F., D. W. Rhind and J. W. Shepherd, *Postcodes: The New Geography*, 1992

Sengoopta, Chandak, *Imprint of the Raj: How Fingerprinting was Born in Colonial India*, 2003

## The Measure of All Things

Clarke, Arthur C., *Greetings, Carbon-Based Bipeds!: Collected Essays 1934-1998*, 1999

Robinson, Andrew and Dipankar Home, 'Tagore and Einstein', appendix to Krishna Dutta and Andrew Robinson, eds, *Selected Letters of Rabindranath Tagore*, 1997

Whitfield, Peter, *The Image of the World: 20 Centuries of World Maps*, 1994

a = above, b = below, l = left, r = right, c = centre

AIP Emilio Segre Visual Archives/Ferdinand Ellerman 68a, 77c; AIP Emilio Segre Visual Archives/Hale Observatories 154a; Antikythera Mechanism Research Project 137a; Art Archive/Egyptian Museum, Turin/Dagli Orti 8b; Ashmolean Museum, Oxford 51; Peter Atkins, *The Periodic Kingdom*, London, 1995 89r; Anthony Aylomamitis 246; Simon Berthon and Andrew Robinson, *The Shape of the World*, London, 1989, 21, 23bl, 136; Bettmann Newsphotos 126; Bildarchiv Preussischer Kulturbesitz, Berlin 123; Bodleian Library, University of Oxford (Douce A.618(16), 19bl; Bridgeman Art Library/Private Collection 35b; British Library, London 9b, 22b, 25a, 52t, 108, 109b, 149l; British Museum, London 2–3, 6, 32, 35a, 48b, 53, 158, 194; Nancy Burson 204; Benjamin Butterworth, *The Growth of Industrial Art*, Washington D.C., 1892 170a; California Institute of Technology 105b; Camera Press/Gamma/Lenhof-Rey 97; CERN, Geneva 82a, 82b; Allan Chapman, *Dividing The Circle*, Chichester, 1995 55, 56a; CSIRO, Australia 96l; R. D. Connor, *The Weights and Measures of England*, London, 1987 60a; Corbis/Historical Picture Archives 52b; Corbis/Bettmann 73ar, 86l, 162a, /NASA 140; Tobias Dantzig, *Number*, New York, 2006 33a; Table of scripts after John DeFrancis, *Visible Speech*, Honolulu, 1989 160; Delambre, *Base 2*, pl. VII, photo Roman Stansberry 28bl; Alexandra Dell i Niella 182al; Derriford Hospital, Plymouth 182r; O. A. W. Dilke, *Mathematics and Measurement*, London, 1987 22a, 54a; E. Dunkin, *The Midnight Sky*, London, 1869 (Astr.8178.3) 23al, / *The Midnight Sky*, London, 1891 68b; Albert Einstein Archive, The Hebrew University of Jerusalem 137b; Empics/Associated Press 125a; Empics/Ifremer, A. Fifis 135r; *Encyclopaedia Britannica* 162b; ESA-P. Carril 110a; M. C. Escher, *Circle Limit III* © 2006 The M. C. Escher Company-Holland. All rights reserved. www.mcescher.com 45; ESO/Rainer Schödel (MPE) et al., NAOS-CONICA 152; Flickr/Herman Yau 31b; Flickr/Jon Delorey 38b; Flickr/Cameron Booth 42a; Flickr/Kim Smith 60bl; Flickr/Brian Aslak Gylte 72a; Flickr/ Betsy Enslin 91a; Flickr/Elizabeth West 91b; Flickr/Rodd Halstead 92bl; Flickr/Stuart Worrall 103a; Flickr/Sarah Jane Rhee Danyluk (www.sarahjanerhee.com) 167bl; Flickr/Álvaro Ibáñez (Microsiervos.com) 208b; Steve Fricker/Folio Art.com 202b; Getty Images 151b; Getty Images/Hulton Archive 44br; Getty Images/Tim Flach 102r; Getty Images/Robert Clare 201al; Getty Images/AFP 209a; GFZ Postdam PR 26a; Nemai Ghosh 210b; William N. Goetzmann and K. Geert Rouwenhorst, *The Origins of Value*, New York, 2005 59; Gordon Gould 77b; *The Graphic*, 8 Aug. 1885 78a; Sonia Halliday Photographs 49l; Anthony Haythornthwaite 113; Sandra Hempel, *The Medical Detective*, London, 2006 192r; E. R. Henry, *Classification and Uses of Finger Prints*, London, 1901–1922 201b; Leofranc Holford-Strevens, *The History of Time*, Oxford, 2005 61, 63r, 197b; Steele Hill and Michael Carlowicz, *The Sun*, New York, 2006 © Vic Winter/ICSTARS Astronomy 148bl; R. Hooke, *Animadversions*, London, 1674, Tabula 1a 56b; David Howard, Department of Electronics, University of York 173; G. Hulbe, *Einwanderer erster und zweiter generation aus Mittel-und Westeuropa*, Stuttgart and Hamburg, 1940 200; *Illustration*, 16 May 1874, engraver H. Dutheil, photo Roman Stansberry 30l; Institut Bruno Comby, Houilles, France 90; Japan Meteorological Agency 115a; Karpeles Museum, Santa Barbara, photo David Karpeles 29bl; W. M. Keck Observatory/Sarah Anderson 67b; Martin Kemp, *Seen/Unseen*, Oxford, 2006 169a; © J. M. Kenoyer, courtesy Dept. of Archaeology and Museums, Govt. of Pakistan 8a; Landesamt für Denkmalpflege und Archäologie Sachsen-Anhalt, photo Juraj Lipták 9a; Simon Lawson, Diamond Trading Company 133a, 133b; Library of Congress, Washington D.C. 18; James Lipton, *An Exaltation of Larks: The Ultimate Edition*, New York, 1991 © Kedakai Lipton 211; The Master and Fellows of Magdalene College, Cambridge 164c; Dennis Mammana/Skycapes 147; E. J. Marey, *La Méthode Graphique*, Paris, 1885 13; C. R. Markham, *Memoir of the India Survey*, London, c. 1870 109a; © Alexander Marshack, *The Roots of Civilization*, Mount Kisco, New York, 1991 34r; Roland Melzack and Patrick D. Wall, *The Challenge of Pain*, London, 1996 191; Metric Association UK 31a, 103b; Peter Michaud (Gemini Observatory), AURA, NSF, 150a; M. L. Design 74bl, 74br, 120; Jeff Moore (jeff@jmal.co.uk) 12a (Tom Tom equipment), 43b, 171b; Philip and Phyllis Morrison and the Office of Charles and Ray Eames, *Powers of Ten*, New York, 1982 1, 149r, 178a, 206ar; Musée de Laon 28br; Musée des Arts et Métiers-CNAM, Paris, photo CNAM 28a; Musée du Louvre, Paris 33b, 40a; Musée National du Château, Malmaison 29br; Museum Boerhaave, Leiden 141a; Museum of the History of Science, Oxford 4; Muslim Heritage.com 63l; NASA 131, 168; NASA/ESA/S. Beckwith (STScI) and The HUDF Team 80; NASA/C. Mayhew and R. Simmon (GSFC) NOAA/NGDC, DMSP 94; NASA/Kennedy Space Center 100; NASA/GSFC, MODIS 106; NASA/University of Iowa 111; NASA/(JPL)/ESA/Italian Space Agency 142a, J. Clarke (Boston University) and Z. Levay (STScI), ESA 142b; NASA/GFC 144al, 144ac, 144ar; NASA/Malin Space Science Systems, MGS, JPL 144b; NASA/K. Gordon (University of Arizona), JPL-Caltech 150b, Stardust Team, JPL 151a; NASA/WMAP Science Team 155a and 155b; National Gallery of Art, Washington DC 34c; National Institute of Standards and Technology, Gaithersburg, Maryland 79; National Library of Scotland 49r; National Maritime Museum, London 23ar, 24a, 24b, 107, 197a; National Physical Laboratory, Teddington 64, 71b, 78b, 85a, 85b, 96r, 101; National Physical

Laboratory, Teddington/Andrew Hanson 99l, 99r; National Portrait Gallery, London 26br; Natural History Museum, London 74al; NCAR, Boulder, Colorado 116bl, 116br; Howard Bluestein, University of Oklahoma, NCAR, Boulder, Colorado 116a; Victor Neiderhoffer 34l; Charles S. Neumann 114; NOAA, Frank Marks 115b; NRC Institute for National Measurement Standards (NRC-INMS) 95a; Oak Ridge National Laboratory, Tennessee 70l, 70ar, 70b; G. Palatino, *Libro Nuovo, c.* 1540 163a; A. Park, *An Apothecary with a Pestle and Mortar to Make up a Prescription,* engraving 189a; Samuel Pepys, *Memoires Relating to the State of the Royal Navy,* London, 1690 164a; Photothèque des Musées de la Ville de Paris. Photo Svartz 29al, Photo: Chevalier 195; Rex Features/SNAP 206br; Andrew Robinson, *Einstein,* New York, 2005 138r; Paul W. K. Rothemund 105a; Royal Naval College, Greenwich 73br; Royal Society, London 175a; Hans-Ulrich Schmincke 132l, 132r; Science & Society Picture Library/Science Museum 10, 54bl, 71a, 73l, 186a, 186b, 198b; Science & Society Picture Library/University Museum of Archaeology & Ethnology, Cambridge, Massachusetts 47; Science Photo Library (SPL) 69b; SPL/Sovereign, ISM 12b, 156; SPL/Gregory Sams 44a; SPL/American Institute of Physics 75b; SPL/National Solar Observatory 76; SPL/Andrew Syred 89l; SPL/Peter Menzel 118l; SPL/BSIP, PIKO 179; SPL/John Cole 188; SPL/Kevin Curtis 180; SPL/Geoff Tompkinson 182bl; SPL/Eye of Science 187a; SPL/Martin Dohrn 202a; SPL/NASA 212; Science Museum, London 50b, 93; Schloss Ambras, Austria 67a; Screwfix 171c; Scripps Institute of Oceanography/UCSD 119a; Chandak Sengoopta, *Imprint of the Raj,* London, 2003 201ar; Dava Sobel and William J. H. Andrewes, *The Illustrated Longitude,* New York, 1998 26bl; SOHO/NASA/ESA 148al, 148ar, 148cl, 148cr; Swedish Royal Academy of Sciences. Photo Georgios Athanasiadis 28ar; Swedish Telescope Institute for Solar Physics 145b; Richard P. Taylor 44bl; Marie Tharp, 124; Vasily Titov (Vasily.Titov@noaa.gov) first published in *Science,* 23 Sept. 2005 128; Tom Tom (screen) 12a; Edward R. Tufte, *The Visual Display of Quantitative Information,* Cheshire, Connecticut, 1983 190b; Utrecht University Museum, Institute of History and Foundations of Science, courtesy Rob van Gent, Tiemen Cocquyt and Carl Koppeschaar 141b; University of Arizona, Tree Ring Research Laboratory 122a; University of Cambridge Library 121; University of Cambridge Library/negative 7857 from RGO.118 62; University of Oklahoma Libraries/History of Science Collections 145a; University of Utrecht 214–15; US Federal Government 175b; Vatican Museums, Rome 209b; Jean Vertut 46; James D. Watson with Andrew Berry, *DNA,* London, 2003 © Wellcome Photo Library/The Sanger Centre 178b; Adrian White, University of Leicester 199; Peter Whitfield, *The Image of the World,* London, 1994 25b; wikipedia.org 69a, 84a, 134; wikipedia.org/Library of Congress,

Washington, D.C. 153; www.worldmapper.org © 2006 SASI Group (University of Sheffield) and Mark Newman (University of Michigan) 207; Yale University Library, Babylonian Collection NBC 7309 57.

The publishers are grateful to the following sources for reference and information in producing the diagrams and artworks: *New Scientist* 16, 87, 143, 180l, 184l and 184r; Ken Alder, *The Measure of All Things,* London, 2002 27; R. D. Connor, *The Weights and Measures of England,* London, 1987 30al; Charles Seife, *Zero: The Biography of a Dangerous Idea,* London, 2000 39; National Physical Laboratory 65a, 65b, 99lr, 110b; J. Smith and N. A. Beresford, *Chernobyl Catastrophe and Consequences,* Chichester, 2005 75a; Peter Atkins, *The Periodic Kingdom,* London, 1995 89; wikipedia.org 91, 165, 171a, 198; Alex Hebra, *Measure for Measure,* Baltimore, 2003 95b, 98 (both); Gavin Pretor-Pinney, *The Cloudspotter's Guide,* London, 2006 113; Felix M. Gradstein, James G. Ogg and Alan G. Smith, *A Geological Time Scale,* Cambridge, 2004 121, 122b; Bruce A. Bolt, *Earthquakes,* New York, 1999 127 (both); Rosaly M. C. Lopes, *The Volcano Adventure Guide,* Cambridge, 2005 129a and 129b; Rainer Flindt, *Amazing Numbers in Biology,* Heidelberg, 2006 135l and 135c; Patrick Moore, *Patrick Moore on the Moon,* London, 2001 139; John Naughton, *A Brief History of the Future,* London, 2000 169b; Jean Pope, *Medical Physics Imaging,* Oxford, 1999 181; Shirley Bond, *Home Measures,* London, 1996 192.

# Index